BUILD YOUR OWN CUSTOMIZED TELESCOPE

BUILD YOUR OWN CUSTOMIZED TELESCOPE

RICHARD F. DALEY AND SALLY DALEY

TAB BOOKS Inc.
Blue Ridge Summit, PA 17214

FIRST EDITION

FIRST PRINTING

Printed in the United States of America

Library of Congress Cataloging in Publication Data

Daley, Richard F., 1948-
Build your own telescope.

Includes index.
1. Telescope—Design and construction. I. Daley, Sally J., 1948- . II. Title.
QB88.D285 1986 522'.2 85-27880
ISBN 0-8306-2656-5 (pbk.)

Contents

Acknowledgments

THANKS TO THE HELP OF RICHARD L. WEINBRENNER OF CyberPak Co., much insight was gained into the skills needed for the computer interface controlling the stepper motors.

We would like to express a debt to D. Eric Allen from Oklahoma. A description of a telescope built by Mr. Allen was published in the Fall, 1984 issue of *Telescope Making* magazine. This coincided with the time we were putting the final touches on the design of the telescope described in this book. Mr. Allen's telescope was remarkably similar to the one we had envisioned and, as a result of careful study of his descriptions and photographs, we made some modifications to the design of our telescope. These modifications helped us to make a much improved telescope.

Introduction

As you consider all the work that must go into building and computerizing a telescope, a natural question for you to ask is, "Why construct my own telescope when I can buy one ready-made?" There are several answers to this question. First is the satisfaction of observing the heavens using a telescope you built yourself. Next is the realization that no commercial telescope can equal the performance of the one you can build yourself without costing thousands of dollars more. Finally, you will have learned the skills with which you can begin a new hobby—designing and building telescopes.

Of course, no project is without its pitfalls and telescope building has one you should consider. The universe, being a very seductive mistress, is continually presenting objects that are a little more distant and a little more difficult to see. Many Amateur Telescope-Makers (ATM) are lured into building bigger telescopes with which to view these elusive wonders. As they build, many of these individuals lose sight of their original purpose—celestial observation—and develop a new goal—building bigger and better telescopes. So as you build, beware of this pitfall.

Our intent is to not only take you step-by-step through the process of building a telescope, but to give you an understanding of the techniques used in telescope construction. Many books on telescope-making present techniques suitable for the advanced Amateur Telescope Maker but leave the beginner on his own to figure out the basic construction techniques for building and then setting up the telescope for use. This book, for example, will teach you how to find your longitude and latitude and how to align your optics.

The telescopes described in this book have been designed for ease of construction, using readily available materials. For those items specific to telescope-making—such as the optics and stepper motors— we've included recommendations for specific suppliers and suggestions on how to find other sources.

This book is divided into two basic parts:

Telescope Construction. By following the directions in this section, you will build a Newtonian reflector telescope with a fork mount. When you finish this section, you can stop there knowing that you have built a completely functional telescope system.

Computer Tracking. Here you will learn how to add a computer-operated dual-axis tracking system to your new telescope using a Commodore 64 computer, disk drive, and video monitor. This system gives you the convenience of automatically following any object as it moves across the sky. You can type in the location of any object in the sky and the telescope will find and track that object.

The type of telescope you are about to build is unique. While many people build telescopes, few use a computer to control them. With the advent of the personal computer, this is now a feasible option for even the beginner in amateur astronomy. The skills required to design such a system include the mechanical skills needed to construct the telescope itself, the electronic skills to build an interface to allow your computer to control the motors to drive the telescope, and finally programming skills for writing the programs to operate the interface.

While the skills needed are several and varied, you can be assured of success by following the detailed step-by-step instructions. Through many hours of toy and furniture building in the workshop, we developed the necessary mechanical skills to build the telescope. Our programming skills were developed through 22 years of programming, eight of which were professional programming for personal computers.

Chapter 1

A Pep Talk

BUILDING AND COMPUTERIZING YOUR OWN TELESCOPE CAN intimidate and discourage you if you allow yourself to be overwhelmed with the scale of the project either before you begin or at some point during construction. The goal of this chapter is to help you gain a sense of perspective about the whole project as well as to help you be prepared to use the telescope when it is completed.

The first step in gaining a sense of perspective is to decide if you have enough time to set aside for the project. For the beginner, it will take approximately 250 hours of work to complete the whole project. Also, if you grind your own mirror, it could take another 200 hours of work, depending upon the size of the mirror you are grinding. If you have your mirror custom ground, you can expect a 3- to 9-month delivery time.

The next step, after determining whether you have enough time available, is to make some decisions about what size telescope to build. Chapter 2 will help you through that process by actually teaching you many of the technical aspects of how a telescope works. With this knowledge, you can be genuinely comfortable with your decision about size.

When all the decisions have been made and you are actually ready to begin work, you need only to follow the step-by-step instructions in the remaining chapters. We have endeavored to make these instructions as easy as possible to follow, and success can be assured by following the directions one step at a time. The only skills you need for the actual construction are the abilities to use some simple tools and to follow directions carefully. If you have these skills, you can succeed.

As the computer programs are already written, you need no more skill with the computer than the ability to set up your equipment, turn on the power, and load a program. You can either order a diskette containing the programs from TAB Books (see the coupon at the back of this book) or you can type in the program listings.

To actually use the telescope, you need only a basic beginning knowledge of the night sky and a willingness to learn. No other special skills are needed. As you spend time observing, your knowledge will grow and you will develop a deeper appreciation of the night sky.

If you are an experienced amateur astronomer, you can skip the next several paragraphs and go on to the next chapter—you already know the things written here. If you are not an experienced amateur astronomer, read these paragraphs and think about them carefully.

During the time you are building your telescope, you might become so involved with this project that you could be spending all of your spare time—and perhaps even some time you can not spare—in the workshop. Enjoy your labors, but don't forget to set aside some time every clear night to spend outside learning your way about the sky. Look at the sky with your naked eyes and, if you have a pair, with binoculars. Learn where some of the prominent features of the sky are and how these features change from night to night. This study will help you to better appreciate the heavens when your telescope sees first light.

If you do not do your homework in learning about the universe, your telescope will be of much less benefit to you. The secrets of the universe will only be buried deeper in the hundreds of new stars now visible in the eyepiece than you could see with your unaided eye. You will likely become discouraged with your telescope and, like thousands of others who use a telescope unprepared, soon decide you have been deceived, feeling that there is nothing of value to you in the sky. By gaining a good general knowledge of what is in the sky that can be seen unaided, you will then be ready to search out the less visible or invisible objects in the sky.

Another word of encouragement: any successful amateur astronomer can tell you stories that will make your discouragement seem small. Stories of dropped eyepieces or mirrors, the airplane that flew through the field of a prize photograph or the hours spent searching for an object in the wrong part of the sky. Be persistant.

Once you have paid your dues, the universe will reward you with its beauty and wonder.

You are now a graduate of the Telescope Maker's Pep Talk, a prerequisite to the rest of the book. If you do not have a "yes" feeling about tackling the project, you better think hard about going further. If you feel a spark of enthusiasm, you are ready to go to work. And remember, the work involved in making a telescope is just the beginning of a lifelong journey among the stars.

Chapter 2

Telescope Size and Site

PURCHASING OR BUILDING A TELESCOPE IS AN EXCITING ADventure. This is true whether it is your first telescope or the one you have spent years planning for to become your ultimate telescope. A quality telescope of any size can bring you and your family years of enjoyment. It also can allow you to engage in serious research in the area of astronomy where you, as an amateur scientist, can actually make a genuine contribution.

The choice of telescope is as important for the beginner as for the experienced observer. Many people make a poor choice for their first telescope, then quickly lose interest in astronomy. No one telescope is perfect for all possible endeavors and to choose the one best for you, there are several things you should consider:

- ***Design***—A telescope designed for viewing the planets is different than one designed for viewing deep space objects.
- ***Optics***—The larger your mirror or lens, the larger your telescope will be and the more it will weigh and cost.
- ***Space***—Do you have an observatory or flat dark spot for observing in your backyard or will you need to travel for some distance each time you want to observe?
- ***Experience***—An inexperienced astronomer can quickly get in over his head with a telescope too big and too powerful. Unless you have a permanent location for your telescope, you must be pre-

pared to lug your telescope to your viewing location each time you want to use it.

- ***Magnification***—The more powerful your telescope, the more difficult it is to find the objects you want to see. It is better to start small and develop your observing expertise then move up to a larger telescope.

Becoming informed in each of these areas will give you a good basis for deciding what size and type telescope is best for you.

There are three main types of telescopes used by both the beginner and the advanced amateur astronomer. They are the *refractor*, the *compound*, or catadioptric, and the *reflector*. Each has its strengths and weaknesses, and this along with different design gives each its own particular best use. While much could be written about each, we will include only enough information to show you why the reflector telescope was chosen for this book.

The refractor is the type of telescope most people visualize when they think of astronomy. See Fig. 2-1 for a drawing of the basic refractor telescope. The refractor has a long tube with an eyepiece at one end and an objective lens at the other. Because of the simplicity of its design, nothing obstructs your view as you look through the eyepiece down the tube and out through the lens. Many people feel they get the best view of the night sky looking through a refractor. The refractor telescope is the telescope of choice for viewing the moon, planets, and sun. But for most amateurs the size and cost of a refractor make owning one prohibitive. The typical refractor owned by the amateur has an objective with a diameter of either 2.4 inches or 4 inches. The tube lengths are about 40 inches and 60 inches long, respectively, with a cost of around $400 and $2150, respectively.

The second type is the compound, or catadioptric, telescope. There are two basic designs of the compound telescope—the Cassegrain telescope and the Maksutov telescope. See Fig. 2-2 for drawings of the basic designs of the Cassegrainian and Maksutov telescopes. The compound telescope is perhaps the best all-around commercial telescope available for amateur use, with the Schmidt-Cassegrain being the most frequently bought. The popularity of the compound telescope is accounted for by its relatively low weight and low cost when compared with a refractor and its advantages

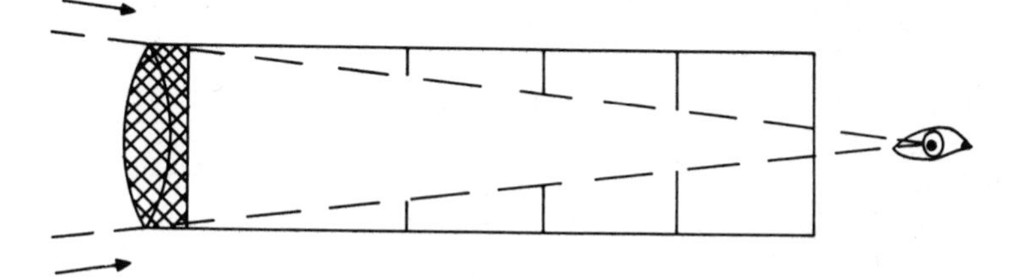

Fig. 2-1. Basic design of a refractor telescope.

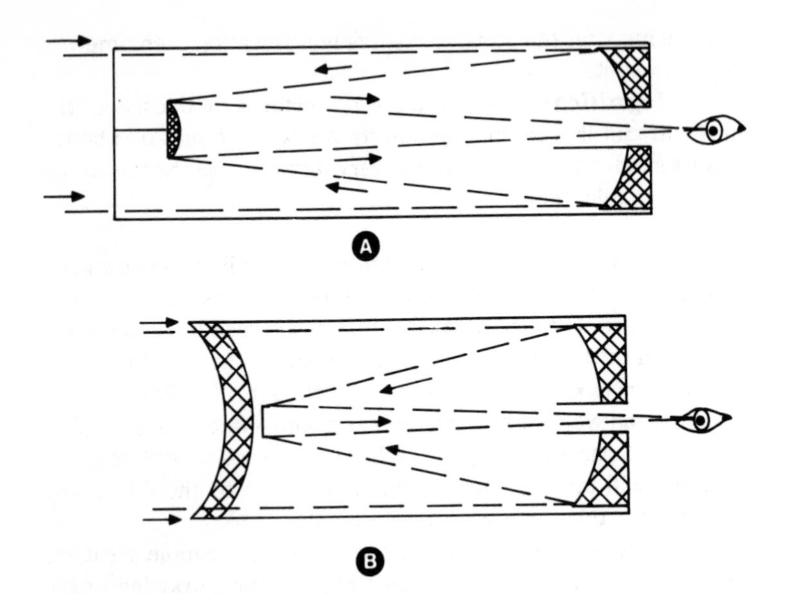

Fig. 2-2(A). Basic design of a Cassegrain telescope. (B) Basic design of a Maksutov telescope.

when used for photography. The greatest disadvantage of the compound telescope is the difficulty of aligning, or collimating, the optics, and then maintaining the alignment. The compound telescope can be used with about equal success for observing either near objects—like the sun, moon, and planets—or deep space objects.

The third type of telescope commonly used by amateurs is the reflector. See Fig. 2-3 for a diagram of the basic reflector telescope. The most common reflector is the Newtonian reflector, which is the type described in this book. The reflector works by having a mirror at one end of the tube, which converges and reflects the light back to another, smaller, mirror at the other end. This mirror, placed at a 45° angle, reflects the light through an eyepiece on the side of the tube. Newtonian reflectors are primarily designed for viewing faint deep space objects. They give fine views of the planets and the moon, but these objects would be so bright in your eyepiece that you may find it difficult to see some details. The reflector is an excellent telescope for the amateur because it is still fairly portable, even with a relatively large mirror. It also is easy to align and the cost of construction is low. Of the three types, the reflector is the ideal one to build because of its low cost and simple design.

If you are new to astronomy, be aware that while many fine commercially made telescopes are available for the amateur, do not be misled by the claims made for small department store telescopes. These often tout excessive magnifications like 454×, 600×, or occasionally even 1000×. While optically possible, these magnifications are impractical for actual observing. We seldom use magnifications above 200 on our telescopes and 200 only infrequently. These department store telescopes are only cheap imitations of real telescopes. They are wobbly, hard to use, and give fuzzy views when you do find something.

Quality telescopes are sized, not by their magnifications, but by the size of their optical system. Telescopes are described as 8 inch (20 centimeter), 10 inch (25 cm.), 12.5 inch (32 cm.), etc. For the reflector, this numbering system refers to the size of the primary mirror. In this book, we describe the construction of 8-inch, 12.5-inch, and 16-inch telescopes. The one you see illustrated in the text is the 12.5-inch telescope. Which one should you build? Let's take a look at the important factors that you'll need to know to make the right decision for you.

The cost of building a telescope is in direct proportion to the size of the telescope. Again, what determines the size of the telescope? The optics. And so it is when building. You can expect the cost of building any of the telescopes described in this book to be about the same for all items you will need except the optics. It will cost about $400 to $500 for materials, eyepiece holder, stepper motors, and controller interface. At this time the Commodore 64 and system will cost about $750. The price for an 8-inch mirror is about $100, a 12.5-inch mirror about $300, and a 16-inch mirror about $750. Depending upon the source for your optics, the cost may vary, but the ratios given above hold true for a number of optics manufacturers.

The next thing to consider is the size-to-weight ratio. Using the construction techniques and materials presented in this book, an 8-inch telescope will weigh about 50 pounds, a 12.5-inch telescope will weigh about 130 pounds and a 16-inch telescope will weigh about 220 pounds. If you have to carry a 220-pound telescope 100 yards—or even 100 feet—from its storage place to your observing site, you may find yourself disinclined to use it very often. And if you can not use it in your backyard, the situation would be even worse.

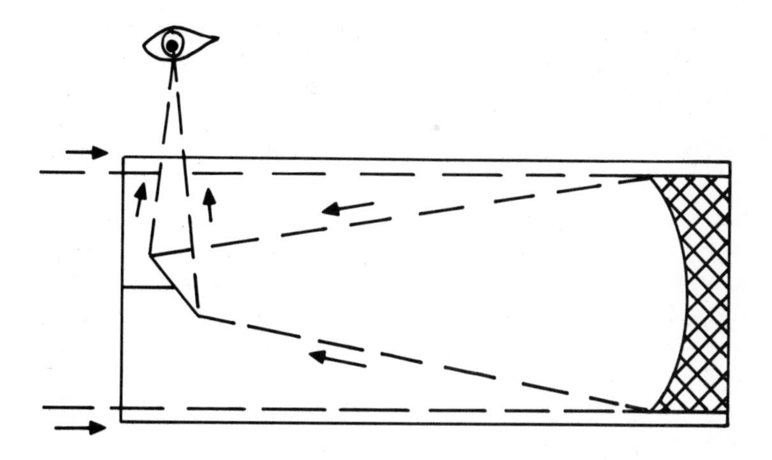

Fig. 2-3. Basic design of a reflector telescope.

Another factor to consider when deciding which size mirror to choose is the performance of each size of telescope. Performance—how much you can see with a particular telescope—is dependent on two points—the size of the optics and how much light is visible from the celestial object under observation. This assumes ideal viewing conditions of little atmospheric turbulance and no wind. A big reflector mirror gathers more light than a small one and it is the light-gathering ability of the telescope—not its magnification—that is the key to successful observation. Stated simply, the larger the mirror the more light a telescope can collect and the more you can see. You will want to choose the largest possible mirror that is in keeping with how much you are willing to spend for optics and how portable the finished telescope must be.

The other factor used to determine just how much you can see with a particular telescope is how much light can be seen on Earth from individual stars, planets, galaxies, etc. How bright a particular celestial object appears to the human eye is known as its *magnitude*. Magnitude usually refers to the relative or apparent magnitude, that is, how much light we are able to see as we look at an object. At times magnitude is referred to as actual or absolute magnitude. This means how much light is actually being emitted from an object—a larger hotter object will emit more light than a smaller cooler object.

With the unaided eye, the faintest stars which you can see are at magnitude 6.5. As the numbers get smaller, the objects become brighter. A magnitude 5 star is 2.512 times brighter than a magnitude 6 star. While it may be a bit confusing, the smaller the magnitude number the brighter the object; the larger the magnitude number the fainter the object. To see the fainter objects in the sky, you would need a telescope with a larger mirror. For example, a 16-inch mirror has four times the surface area of an 8-inch mirror and gathers four times the amount of light. You can expect to see stars with a 16-inch telescope that are four times fainter than the faintest star you can see with the 8-inch telescope. It is possible to determine the limit of the magnitude of the stars that you can see with a particular telescope. Table 2-1 lists this limiting magnitude for each of the telescopes described in this book. These magnitude numbers assume ideal viewing conditions.

It may seem paradoxical, but with a larger mirror you actually see a smaller section of the sky than you do using a smaller mirror. At the same time, however, you see more details of what you are observing with the larger mirror. It is a little like studying a flower with a magnifying glass. The more powerful the magnifying glass, the smaller the section of the flower you see, but you see more detail in the flower.

The amount of detail seen in a telescope is called its *resolution*.

Aperature (inches)	Limiting Magnitude
eye	6.5
8	13.9
12.5	14.7
16	15.5

Table 2-1. Limiting Visual Magnitude of Various Telescopes.

The resolution of a telescope is related to the size of the optics as well as the quality of the optics and the conditions under which they are used. This theoretical limit, called *Dawes limit,* assumes perfect optics and ideal observing conditions. Resolution can be calculated by the formula:

$$R = 4.65/d$$

where R is the resolution in arc seconds and d is the diameter of the mirror in inches. One arc second is about the size of a quarter at 100 feet. Table 2-2 shows the theoretical resolution of the three telescopes being considered in this book.

While we have talked a great deal about *magnification,* we have not yet defined the term. Magnification is an apparent enlarging of the image as it passes through the eyepiece. When you increase the magnification, you are looking at a smaller portion of the image produced by the mirror and seeing it in greater detail. The amount of magnification you can use with a particular telescope depends upon the size of the mirror, the resolution, and the viewing conditions of the sky. A telescope with a large mirror, good resolution, and excellent viewing conditions is capable of increasing the magnification toward the maximum limit of that telescope system.

One other limiting factor for magnification is the eye. Under dark sky conditions, with the human eye adapted to the dark, the pupil usually measures about 8 millimeter, or roughly 1/3 of an inch, in diameter. A beam of light larger than this emerging from the eyepiece would result in *vignetting* (loss of some of the image). The lower magnification eyepieces give the larger light beam sizes. Thus the lowest practical magnification given by an eyepiece on your telescope should produce an 8-mm image. The eye has the best reso-

Table 2-2. Resolution of Telescopes According to Dawes Limit.

Aperature (inches)	Resolution (arc seconds)
8	0.57
12.5	0.36
16	0.29

lution when the pupil is about 3 mm. An eyepiece then that produces an image of 3 mm is the best, or optimum, magnification to use. The higher the power of the eyepiece, the smaller the image that it produces. If the image is less than 1 mm, it is very hard for the eye to gain much information from it. That then is considered the maximum practical limit of an eyepiece for a telescope.

The size of the image produced by the eyepiece is called the *exit pupil.* The exit pupil can be calculated by the formula:

$$E = A/M$$

where E is the exit pupil, A is the aperture, or diameter of the mirror, in millimeters and M is the magnification. Table 2-3 shows these values for the three telescopes in this book.

A final consideration when deciding which size telescope to build is your observation location. There is one major criterion for your observation location—you must have dark skies. Unfortunately, few of us have the luxury of ideal skies from our backyard. Usually, street lights, neon advertising signs, traffic, and many other sources of light pollution interfere with viewing. While various observing filters can help tone down light pollution—allowing us to do some observation from our backyards—most of us must travel to a rural location to find really dark skies. For some observers, this can mean driving 50 miles or more to set up the telescope. The decision on which sized telescope to build hinges on whether you have to travel like this. You may find it hard to transport a 16-inch telescope system in a compact car.

An ideal arrangement is to have the telescope permanently set up in an observatory building in your backyard. A desk for your star charts and notebook, and you are ready to go! This arrangement would allow you to simply walk to the observatory and begin observing.

Few of us have a location for such a building, even if we could afford it, so most of us will have to compromise. At the least, you will need a level location for your telescope. To use the computer, you must also have a source of electrical power. This could be house current or even a car battery and converter. And the widest, darkest view of the heavens you can find. If this is on your prop-

Table 2-3. Minimum, Optimum, and Maximum Magnifications for a Given Telescope.

Aperature (inches)	(mm)	Magnification (Minimum)	(Optimum)	(Maximum)
8	205	25	70	205
12.5	320	40	105	320
16	410	50	140	410

consider pouring a level concrete pad and bringing power from the house to the concrete. If it is not on your property, it would be wise to talk with the owner before you do anything.

Give careful thought when finding a place to observe. It would be very discouraging to prepare a place, then set up your telescope only to find that your neighbor has a tree that blocks the best section of the sky. Before you do anything permanent, spend several nights with the telescope at the proposed location. Make sure you have the best place for you.

Finally, it is important to get the exact location of your observing site. When you decide to contribute research results, you will need to have the latitude, longitude and altitude of your observing location. You will also need to know the latitude of your observation for building portions of your telescope. A number of sophisticated methods exists for doing this, but the easiest way is to pinpoint your position on a topographic map supplied by the U.S. Geological Survey. These maps are available for any location within the United States. Your local library will have copies of these maps or they can be purchased from a surveyor's office or the outdoor section of a sporting goods store or direct from the U.S. Geological Survey office.

Which telescope for you? Here are our recommendations. Unless you are an experienced and dedicated amateur astronomer with a permanent observing location, postpone the 16-inch telescope until later. The cost and hazards of transporting it are too great. We began observing with a 4-inch Newtonian reflector which is easy to pick up, tuck under your arm, and go. In a short time we outgrew the 4-inch, and obtained both a 6-inch and 10-inch Newtonian. The weights there are about 40 pounds and 110 pounds, respectively. Neither scope has a permanent location, and that 110 pounds sure makes it difficult to transport it outside when we are tired or only have a short time for observing. This is when we like the 6-inch telescope. Of course, we have also become addicted to the 10-inch views, making the 6-inch more unattractive for observing.

Avoid the common mistake made by many beginning amateurs: do not construct a telescope larger than you would be willing to move around and so powerful you will have difficulty locating the stars you want to see. If you have never owned a telescope before, build the 8-inch version. With this size you can easily move it about and get a feel for observing the stars. If you find that astronomy is for you, then build a larger telescope. Of course, the decision is up to you. Just watch that you do not go for a telescope too large for your strength and level of expertise. Be sure, however, that any telescope you do own is a solid, stable performer. Build one like ours—or modify it to suit yourself.

Chapter 3

Building the Telescope Mount

A REASONABLE ASSUMPTION THAT COULD BE MADE ABOUT telescope building is that there are as many variations in the design for the construction of the telescope mount as there are builders of telescope mounts. If this is your first excursion into the construction of a computer-controlled telescope, you should carefully follow the design given in this book. However, if you are like most people, you will no doubt see as construction progresses some ways that are better than ours. Keep notes on your ideas and try them on your next telescope.

There are two basic types of telescope mounts—the altazimuth mount and the equatorial mount. Within these two basic categories there are numerous subtypes, and many people who build telescopes vary these subtypes even further to fit their individual needs, equipment, and building ability. Of the wide variety of mount designs to choose from, we have selected a subtype of the equatorial mount called the *fork mount*. Because of its stability, the fork mount is used by most major observatories for their largest telescopes, and it is one of the few mounts that can be completely constructed from wood using common woodworking tools and techniques.

In this and following chapters, we recommend that, before you begin work, you carefully read the entire chapter and visualize each step as we describe it. Often, taking the time to comprehend the entire picture helps the step-by-step instructions to make more sense.

You will find that most of the work on the fork mount can be accomplished with hand tools. In a few cases, a table saw or radial-arm saw would make your work much easier and, while not required, a drill press would also be helpful. A drill press is especially handy if you elect to cut weight-reducing holes in your mount. Otherwise, a circular saw, portable drill, sander, hammer, pliers, and screwdrivers are all the tools required for the completion of this project. Table 3-1 is a guide to the purchase of materials you will need from the hardware store or lumberyard for the different sizes of telescopes.

We belong to the "glue-and-screw" school of woodworking, and while this is somewhat more costly and time consuming than the nail-it crowd would like, the finished product is sturdier and longer lasting. You may choose either method for most joints, but be aware that nails used in the edge of a section of plywood produce a very weak joint.

Table 3-1 lists the material needed to build either an 8-inch, 12.5-inch, or 16-inch telescope. The numbers in parentheses refer to the ten numbered notes at the bottom of Table 3-1. These notes tell how or where you can obtain the necessary parts.

Table 3-1. Materials Required for Building a Telescope.

Item to be purchased	Telescope size		
	8″	12.5″	16″
A-A or A-B grade plywood sheets (1)	1	1	1
A-C grade plywood sheets (1)	2	2	2
1 × 2 lumber 8′ lengths (2)	2	2	2
2 × 4 lumber 8′ lengths (2)	8 (3)	10 (3)	6 (3)
2 × 4 lumber 10′ lengths (2)	0	0	4 (3)
2 × 6 lumber 8′ lengths (2)	1	0	0
2 × 8 lumber 8′ lengths (2)	0	1	1
Dowel pegs 1/4″ dia. by 1 1/2″ long	24	16	16
Dowel pegs 3/8″ dia. by 1 1/2″ long	0	12	12
Steel casters 1 1/2″ (4)	2	0	0
Steel casters 2″ (4)	0	2	2
Teflon plate -1/4″ thick (5)	1	1	1
Aluminum strip 1″ wide 1/8″ thick	4′	6′	8′
Sheet metal strips 1.5″ wide by 24′ long	2	2	2
Roll pins 1/16″ × 3/4″ long	2	2	2
Threaded rod 1/4-20	2′	2′	2′
Threaded rod 5/16-18	3′	3′	3′
Bolts 1/4-20 3″ long	3	3	3
Nylon bolts 1/4-20 1/2″ long w/nuts	3	3	3
Nuts 1/4-20	24	24	24

Wing nut 1/4-20	1	1	1
Nuts 5/16-18	6	6	6
Wing nuts 5/16-18	6	6	6
Bolts 3/8-16 1 1/2″ long	36	36	36
Nuts 3/8-16	36	36	36
Machine screws 10-32 1″ long	3	3	3
Machine screws 10-32 3 1/2″ long	3	3	3
Cap nuts 10-32	3	3	3
Hex head set screw 3/16″ fine	6	0	0
Hex head set screw 1/4″ fine	0	6	6
Threaded inserts 10-32	12	12	12
Threaded inserts 1/4-20	8	8	8
Tee nuts 10-32	6	6	6
Electrical metallic tubing 3/4″ dia.	1′	1′	1′
Electrical metallic tubing 1″ dia.	30′		
Electrical metallic tubing 1 1/2″ dia.		40′	
Electrical metallic tubing 2″ dia.			50′
Pipe flanges (6) 1″	3		
Pipe flanges (6) 1 1/2″	0	3	
Pipe flanges (6) 2″	0	0	3
Pipe thread to EMT adapter to fit pipe flange	1	1	1
Fan belt material	as required (see Chapter 6)		
Rubber stops from sliding glass door	2	2	2
Cabinet door handles	2	2	2
Wood glue (7)	1 pt.	1 qt.	1 qt.
Wood filler like plastic wood	1 can	1 can	1 can
Wood screws—zinc plate 3/4 × 8	75	75	75
Wood screws—zinc plate 1 1/4 × 8 Pan head	4	4	4
Wood screws—zinc plate 1 1/4 × 8	100	150	150
Wood screws—zinc plate 1 3/4″ × 10	50	50	50
Wood screws—zinc plate 2″ × 10	0	50	50
Wood screws—zinc plate 2 1/2″ × 12	25	50	50
Wood screws—zinc plate 3 1/2″ × 12	0	12	12
Small box of 4d finishing nails			
Small box of 1″ brads			
Mirror 1 1/4″ thick f/5 (8) 1 ea. dia.	8″	12.5″	16″
Secondary mirror (8) 1 ea.	2.6″	3.1″	4.25″
Stepper Motor controller HS-3 (9)	1	1	1
Stepper Motors - Clifton #23 (9)	2	2	2
Focuser (10)	1	1	1

1—Obtain exterior grade 3/4″ plywood.

2—Select boards that are straight and knot free. This could mean paying a little more, but it is necessary.

3—If available, substitute one 4-×-6 for three of the 2-×-4s for the 8-inch telescope or one 4-×-8 for four of the 2-×-4s for the 12.5-inch or 16-inch telescopes.

4—Simple steel casters are sufficient for the 8-inch and 12.5-inch telescopes; however, we recommend using ball-bearing casters for the 16-inch telescope. Be sure that the casters are of the flat-wheel variety.

5—Teflon plate can be obtained from your local supplier of plastics.

If they are unable or unwilling to fill a small order, check with Easy-Ware, P.O. Box 247, Darby, MT 59829.

6—There are various lengths of pipe flanges. Choose the 3/4" length.

7—Use any waterproof aliphatic-resin wood glue. We prefer Franklin Industries Titebond glue.

8—We ordered optics from Telescopics, Inc., P.O. Box 98, La Canada, CA 91011.

9—We obtained these items from CyberPak Co., P.O. Box 38, Brookfield, IL 60513.

10—We used a helical focuser. It is more expensive than many others, but much easier to focus. This was ordered from Lumicon, 2111 Research Dr. #5, Livermore, CA 94550.

Be sure that you order the optics for your telescope as early as possible. The dimensions for the telescopes in this book assume a mirror with a focal ratio of f/5. If you choose another focal ratio, you will need to adjust your telescope dimensions accordingly. There are a number of suppliers of quality optics. No matter who you order from, prepare to be patient and allow 4 to 6 months for them to make the mirror. You can buy ready-made optics, but custom-made optics are usually of a higher quality than the off-the-shelf variety.

An alternative to buying a mirror is making your own. Purchasing the glass blank and materials for grinding is much cheaper than buying a finished mirror, and many amateur telescope-makers do undertake the task of grinding and finishing their own mirrors. This job can be challenging and time consuming—the typical mirror takes many hours to grind and polish starting with a flat sheet of glass and the finished product must be optically precise—but making your own high-quality mirror is very satisfying. If you have patience and more time than money or would like to do it for the experience, you might consider this route. However, do not begin with a large mirror. The first mirror you grind should be a 4-inch to 6-inch mirror. If you are successful with a small mirror, then you are ready to tackle a larger one. Unless you are usually adept at mirror grinding, you should increase the mirror size in three or four steps before you start work on a mirror larger than 12 inches.

CONSTRUCTING THE BASE

Your first task in building the telescope mount is to start with the construction of the base, a U-shaped unit with a cross brace added to make it more rigid. Remember, as you build, that your mount must be as rigid and stable as possible to minimize any unwanted movement of the telescope during use. See Fig. 3-1 for a drawing of the U-shaped portion of the base. This section will instruct you in making the arms of the unit.

If you did not obtain 4- × -6 lumber for the 8-inch telescope or 4- × -8 lumber for the 12.5-inch or 16-inch telescopes, you should

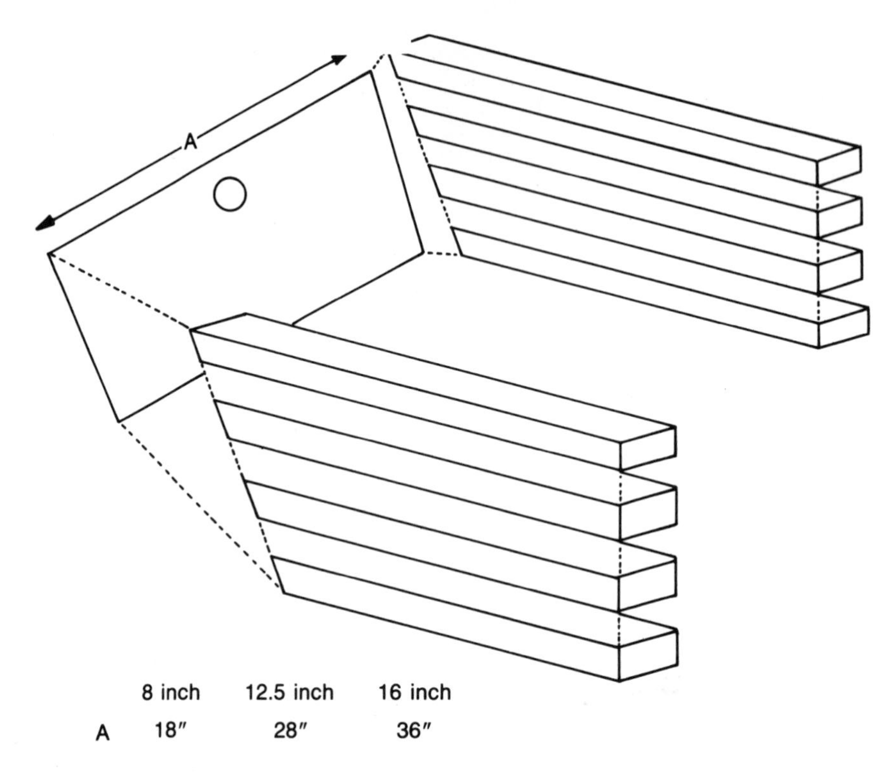

Fig. 3-1. U-shaped portion of the base.

start by laminating four (three, if you are making the 8-inch telescope) of the 2- × -4s together. It is important that your lumber be straight and unwarped. If you are making the 16-inch telescope, use the 10-foot lengths of 2- × -4s. If you are making an 8-inch or 12.5 inch telescope, use the 8-foot 2- × -4s. Glue them with the 4-inch sides together and clamp them tightly so that they form a tight bond. Glue in steps, as it is easier to manage your lumber if you have only one wet glue joint at a time to clamp.

Have you found your latitude, as was suggested in Chapter 2? If not, you should do so while waiting for the laminated 2- × -4s to dry. When the 2- × -4s are dry, or now, if you are using 4- × -6s or 4- × -8s, you will need to know your latitude in order to accurately cut the arms for the base. The arms must be cut according to your latitude, and your latitude may be different than ours. Find the latitude to the nearest degree. Greater accuracy is actually desirable, but it would be difficult to cut the wood with greater precision than this. Greater accuracy can be obtained with the adjustments which are made later.

Once you have your latitude and the lamination of the 2- × -4s is complete, you are ready to begin cutting the wood for the base. In Fig. 3-2, you will see that these laminated pieces need to be cut at an angle on one end and square at the other. This angle of cut is equal to your latitude. It must be cut carefully and both arms of the U-shape need to have exactly the same angle. Even a slight difference will make your telescope difficult to use later. It is here that you will find a radial-arm saw or table saw a definite aid in making sure these two cuts are identical. Because the angle is so crucial to the performance of your telescope, cut it first—one on each side of the laminated piece. Next cut the square end. Figure 3-2 gives you the proper length to cut the arms. Measure the length of each arm, beginning at the apex of the angle. Check again with Fig. 3-2 to insure you have measured properly before cutting. When the arms have been completed, you are ready to construct the bottom of the U-shape.

MAKING THE SOUTH BEARING

The bottom of the U-shaped unit is called the *south end* of the base, because when you use the telescope, it always faces south. This end contains a bearing, which logically enough is called the *south bearing*. To make the south end, select the correct length of cut from "A" in Fig. 3-1 and, using either the 2- × -6 or 2- × -8, depending on which telescope you are constructing, cut it making the ends square. Once you have cut the south end, prepare the south bearing.

Since most of the force on this bearing will be upwards, you will cut out an adjustable bearing pad on the bottom of the bearing. The bearing is lined with Teflon to reduce friction. The south bearing is detailed in Fig. 3-3. First cut a hole in the center of the board so that the edge of the hole is 1 1/2-inch from one side. The diameter of the hole should be 1/2-inch greater than the diameter of the electrical metallic tubing (EMT) you purchased for your telescope. For example, the 12.5-inch scope calls for 1 1/2-inch EMT. Thus the hole cut here would be 2 inches.

Next drill two 5/16-inch holes in from the 2-inch edge of the board closest to the hole you have just cut. See Fig. 3-3 for placement information. These holes should be at least 3- to 4-inches deep.

Once these holes have been drilled, lay out the rectangular cut shown in Fig. 3-3. This rectangular block is the adjustable bearing pad. Be sure that the top edge of the bearing pad runs exactly across the center of the hole you have just cut and that the hole is in the center of the edge of the bearing pad. Carefully drill two holes just

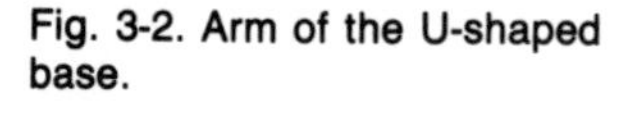
Fig. 3-2. Arm of the U-shaped base.

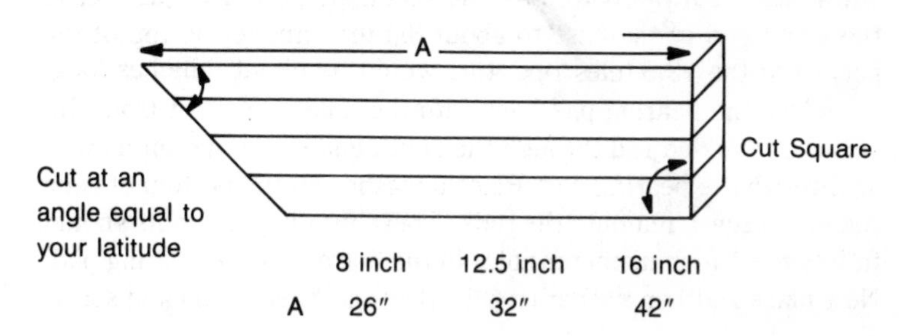

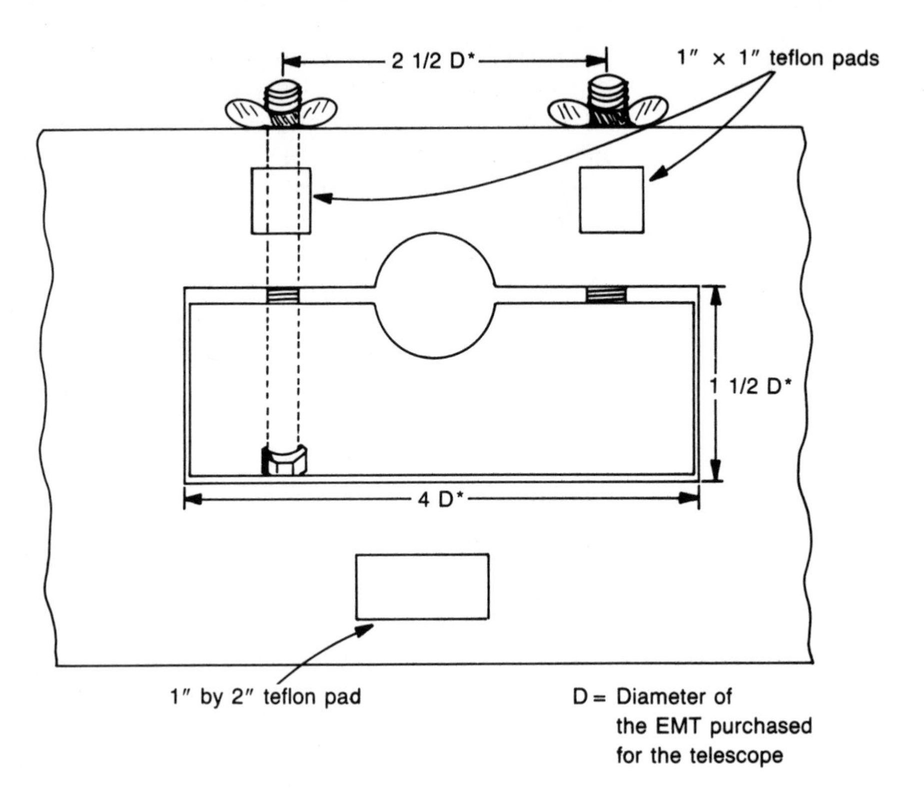

Fig. 3-3. Detail of south bearing on U-shaped base.

large enough for your saw blade in the corners farthest from the bearing hole, placing the holes as close to the lines as possible. If your holes are 3/8-inch or less, they will be cut off later.

Now carefully cut out the bearing pad. Your cuts must be straight since the pad must move smoothly relative to the base of your telescope. Trim about 3/8-inch, or just enough to remove the holes drilled in the corners, from the lower side of the bearing pad, and about 1/4-inch from the upper side.

The 5/16-inch holes, which extend into the block, probably are not completely drilled through the block. Finish drilling the holes through the block now. Then bore two 1-inch diameter holes centered upon the 5/16-inch holes, about a 1/2-inch deep, into the bottom side of the bearing pad.

Cut two lengths of the 5/16-18 threaded rod into lengths that will reach from the bottom of the rectangle you have just cut in the south end of the base to about 3/4-inch above the top of the base. For the 12.5-telescope, this would be about 6-inches long.

Place the bearing pad back into the hole you cut it from. Insert the threaded rod through the south end of the mount as well as through the bearing pad. Place a washer on the bottom of each rod and screw a nut onto the rods. These washers and nuts should fit into the 1-inch diameter holes in the bottom of the bearing pad. Now place another washer on the other end of each rod and screw on a wing nut. Insert a length of the EMT into the bearing, and you are ready to form the Teflon pieces which fit into the bearing.

Cut three pieces of Teflon plate 1 1/2-inch in width and the diameter of your EMT in length. These are to be mounted on the inside surface of the circular hole you drilled into the south end of the base, but first they will have to be formed to fit the inside of the curve. Carefully heat the pieces with a propane torch. When you heat them, be careful. Hot Teflon does not look any different than cold Teflon. When the Teflon is hot, place it between the EMT and the circular hole and tighten the wing nut. Clamping the Teflon into the bearing with the EMT for a few days will help to shape the Teflon.

Once you have formed the Teflon pieces, they are ready to be attached to the wood. Attach them now, or if you are going to paint, or in some other way finish, your telescope, attach them when the paint has dried. If you plan to paint your base, check the fit of the bearing now. With the Teflon pieces in place, tighten the wing nuts. If you have done a good job, you will have a tight fit between the Teflon and the EMT. To attach the Teflon, use small brads, or finishing nails, driving the nail heads below the surface of the Teflon with a nail set. Since the bearing pad will have very little weight on it, mount one piece of the Teflon in the center of the hole in it. The other two pieces should be mounted in the top half of the bearing hole. When the bearing fits properly and the Teflon pieces are in place, fill the holes around the nuts with a wood filler and allow it to dry.

Next cut two pieces of Teflon 1-inch square and another piece 1-×-2 inches. Attach these to the side of the south end that will face to the inside of the U-shaped base, as shown in Fig. 3-3. As before, use finishing nails and drive the ends slightly below the surface of the Teflon, making sure the heads do not protrude above the surface of the Teflon.

Finally, attach this assembly to the angled ends of the arms of the base. Place the parts so that the longest side of the arms are up, and the side containing the wing nuts is up. Fit the pieces together and drill pilot holes for the 2 1/2-inch #12 screws, using at least six screws for each arm. When the pilot holes are drilled, spread a generous layer of glue on the joints and screw them together.

Set this assembly aside while you make the polar cone. The assembly is not finished, but you will need to have the polar cone completed before continuing.

MAKING THE POLAR CONE

The earth is continually spinning from west to east on an imaginary axis, and because of this, the celestial objects seem to be moving from east to west. This is seen most obviously by watching the sun, as it seems to rise each morning in the east and set each evening in the west, but this movement is quickly observed with even the more distant objects when you look at them through a telescope.

As you look at objects through a telescope, they seem to be moving and will rapidly move out of the view of the eyepiece. To watch an object for any period of time, you must frequently adjust the position of the telescope either manually or mechanically. It is with the *polar cone* that you make that adjustment. The axis of the polar cone must be parallel to the axis of the earth—that is why you must know your latitude—and then by using the computer, it can be turned continually in the opposite direction of the earth, and so keep an object continuously in view.

The polar cone consists of two plywood disks with six, (for the 8-inch telescope) or eight (for the 12.5-inch or 16-inch telescope) connecting and supporting ribs. Passing through the center of the disks and ribs is a piece of EMT—the telescope's axis. Figure 3-4 indicates the dimensions and positions of the parts of the cone.

First cut the plywood disks using A-C plywood. While they can be cut with a hand saw, the larger disk must be exact, so a radial-arm saw or table saw is best to use. The instructions here are based on using a radial-arm saw. Experiment with a similar technique if you have access to a table saw.

Cut two plywood squares 1-inch larger than the diameter of the plywood circle and drill a hole into the center of each square. See "A" in Fig. 3-4 for the size of the disk. You will bolt the plywood to the table so that it can be pivoted under the saw, and this hole is the pivot point. To determine where to drill a hole in the table to bolt down the plywood, find a position for the saw blade on the arm such that there will be enough room to freely turn the plywood square underneath it without bumping the support column. Lock your blade in that position. Next carefully measure from the

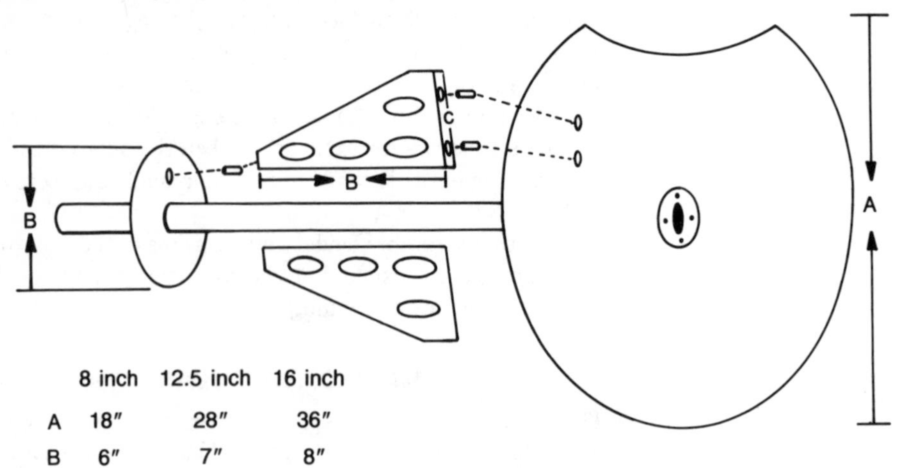

	8 inch	12.5 inch	16 inch
A	18″	28″	36″
B	6″	7″	8″

Fig. 3-4. Exploded view of the polar cone.

Fig. 3-5. Photograph of setup for cutting large plywood disks.

blade the length of the radius—half the diameter—of the disk that you want to cut. We also added 1/16-inch to this measure to allow for sanding. Drill a hole in the saw table at this point the same size as the hole in the plywood squares.

Raise the saw blade, and bolt the plywood square to the table. Tighten it slightly, leaving only enough play to be able to freely turn the square of plywood. Before turning on the saw, double check to make sure that from the point where the blade touches the plywood to the center of the plywood is half the diameter of the desired disk. If it is, turn on the saw and lower the blade enough to cut about 1/16-inch into the plywood. Figure 3-5 is a photo of the setup. Slowly turn the plywood in a complete circle into the direction of blade rotation. After completion of the circle, lower the blade another 1/16-inch and make another cut.

Continue making cuts this way until you have cut completely through the plywood. Do not rush it. Move slowly and carefully through each cut. If you cut too fast, you can cause splintering of the plywood. Once you have finished the cuts for the two large disks, thoroughly coat the "C" sides of the plywood with glue and tightly clamp them together. Allow the glue to dry.

While drying, use a similar technique to cut the small disks for the other end of the cone. An alternative method would be to use a fly cutter on a drill press to cut these smaller disks. The 8-inch telescope only needs one disk, while the 12.5-inch and 16-inch telescopes need two. If you cut two small disks, glue and clamp them together as well.

When the glue on the disks is dry, carefully fill all voids with wood filler. If you have large voids, do not try to fill them all at once. Instead, gradually fill them with several smaller layers.

Using contact cement, attach a piece of plastic laminate to one face of the small disk. For the 8-inch telescope, glue the laminate to the "C" side of the disk. For the other two sizes, glue it to either side. Clamp or weight the laminate for about one to two hours after attaching it. Trim it to the edge of the disk.

Now you are ready to sand the edges of the disks. You can use a belt sander or make a jig out of a hand-held belt sander. Again bolt the plywood disks to a table, using the center pivot hole. Clamp the sander on its jig adjacent to the disk and turn it on. Move the sander to barely touch the edge of the disk and slowly turn the disk into the direction of the motion of the sanding belt. See Fig. 3-6 for a photograph of the setup. Hold the disk firmly as it will tend to rotate in the direction opposite to the one which you are turning. Slowly turn the disk until the edge is smooth, moving the sander slightly closer to the center of the disk as necessary. Work very carefully because it is necessary to make the larger disk as close to circular as possible.

When the telescope is being used, the larger disk will face north, while the smaller disk will face south. The north disk is too large to permit good access to the southern horizon so a notch must be cut into it. If your latitude is less than about 25 degrees, you probably can skip this step. Since we are at 46 degrees, the telescope will not reach closer than 25-30 degrees from the southern horizon and the notch is necessary.

There are a number of methods to cut this notch. You could simply draw an arc and then using a jig saw or band saw, cut it out. The method we chose was one using our radial-arm saw which left an attractive bevel-edged arc. First raise the saw so that it clears the disk to be notched. Then position the saw a distance from the center of the support column equal to the radius of your disk. Carefully draw an arc on the disk; ours was about 6-inches deep. This allows us to view within 12 degrees of the southern horizon. If you are farther south, your notch should be shallower, or farther north, it should be deeper.

Fig. 3-6. Sanding the large disks.

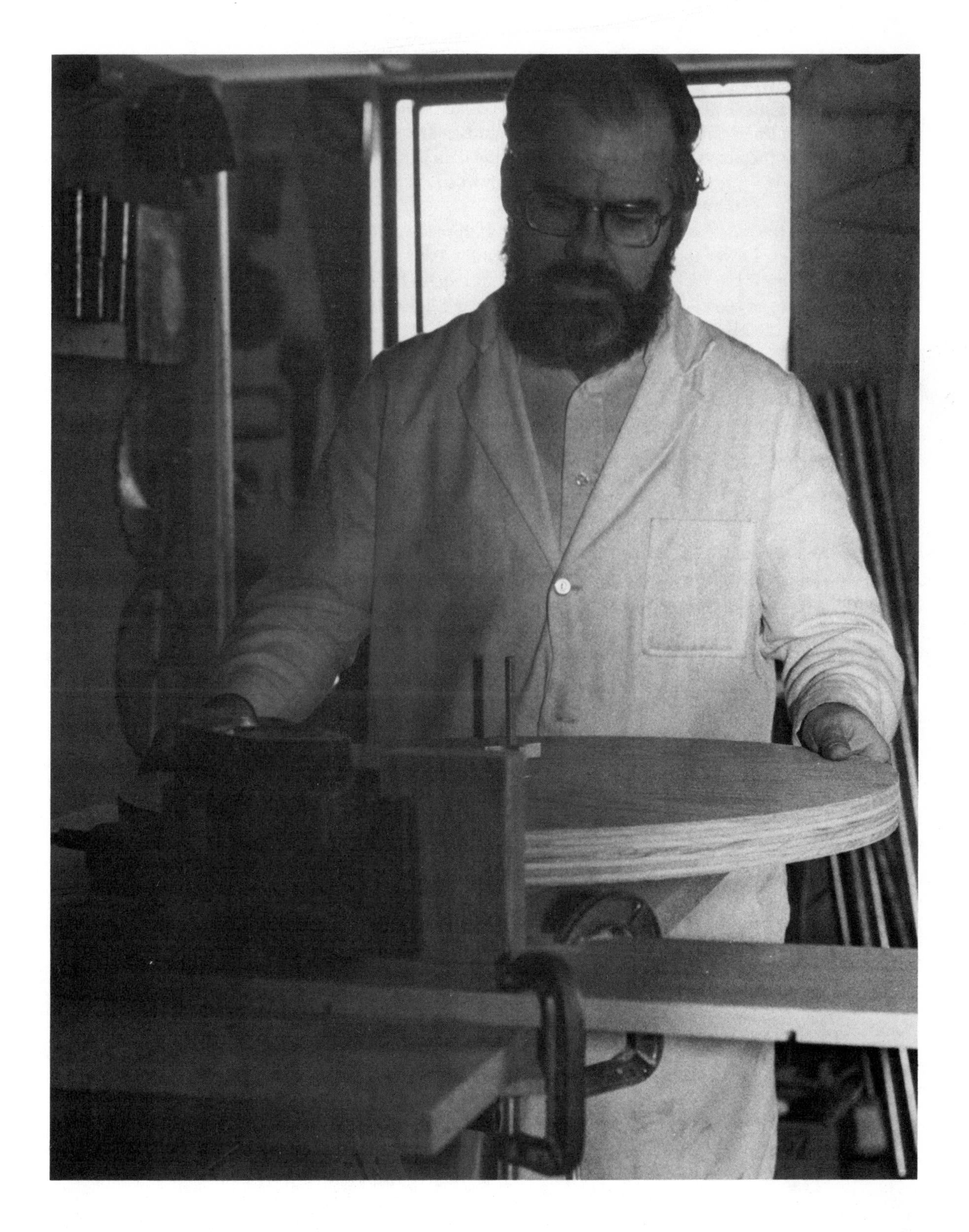

After drawing the arc on your disk, position the disk on the saw table so that when you swing the arm of the saw from left to right, the center of the blade will trace across the arc. Place several sections of 1-inch thick scraps of wood under the disk. Be sure that the area between the arc you drew and the edge of the disk is fully supported. This will prevent splintering of your plywood. When the disk is stable, clamp it firmly to the saw table.

Next, move the saw closer to the column, lock it in place and turn on the power. Lower the saw until the blade will make a cut about 1/16-inch into the disk. Slowly swing it from side-to-side across the surface of the disk. Lower the blade another 1/16-inch and again swing the blade across the surface of the disk. Keep up this process until you have cut all the way through the disk. Then move the blade 1/16-inch closer to the arc you drew on the surface of the disk and repeat the process until you have cut away as much of the disk as desired.

Now drill a hole in the center of each disk. The hole in the smaller disk should be equal in diameter to the outside diameter of the EMT used for your telescope, and the hole in the larger disk should be equal in size to the outside diameter of the EMT adapter. Next rout out an area around the hole in one side of the larger disk to inset the pipe flange. Prepare a template in a scrap of plywood by making a circle the size of the diameter of the pipe flange plus the diameter of the router base minus half the diameter of the router bit. Save this template because you will need to use it two more times. Carefully center the template over the hole in the larger disk, clamp it down, and rout out an area the depth of the lip of the pipe flange.

If you plan to paint or finish your telescope, paint the disks now. (See Chapter 5 for details on painting.) Paint the side of the large disk with the routed out area a flat black to absorb any light that strikes it. Paint the rest any color you choose.

Cut a length of the EMT equal to the length of the ribs plus one foot and attach one end to the adapter. Attach the pipe flange to the larger disk using 1 1/4-inch pan head wood screws. Then screw the adapter into the flange so that the EMT is perpendicular to the disk. This is important—the EMT must be perpendicular to the disk. See Fig. 3-7.

Next cut eight ribs (six for the 8-inch telescope) from the A-A or A-B plywood shaped like those in Fig. 3-8. We cut holes in all but one of these pieces to reduce the weight of the final telescope. When cutting your holes, do not cut them closer than 1 inch from an edge. These holes are optional, but did reduce the weight by about four pounds. The last rib supports the disk in the center of the notch. This rib is trimmed in a line parallel to side "D." Cut off enough so that at least 1/2-inch space will be between the edge

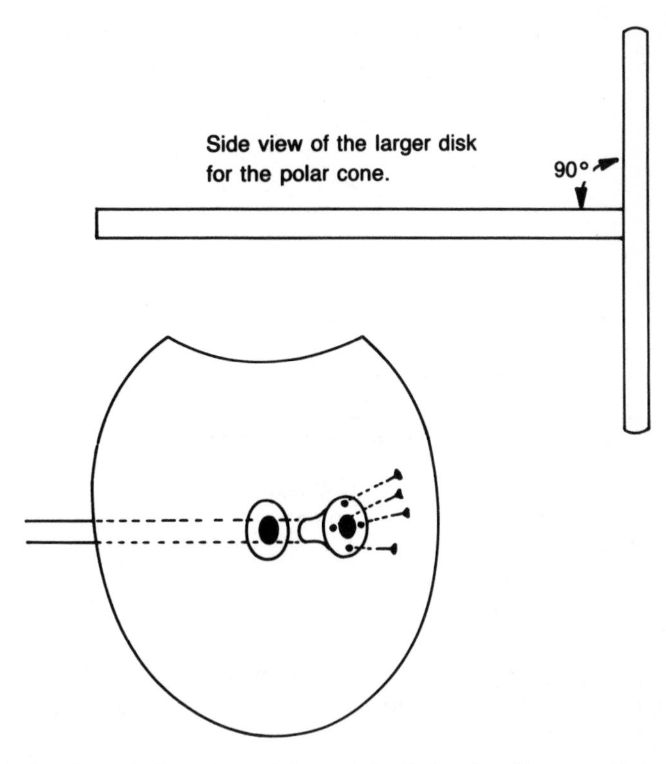

Fig. 3-7. Attaching the EMT.

of the rib and the edge of the notch. Paint the ribs now. It is not necessary to paint the "A" and "C" edges.

The ribs are joined to the disks with dowel joints. Use two dowels on the side labeled "C" in Fig. 3-4 and one in the center of side "A." Equally space them about the disk with a 4-inch space between opposite ribs. The best approach is to attach all of the ribs to the north, or larger, disk and then attach the south, or smaller, disk to this assembly before the glue dries. That way you can make any necessary adjustments to the ribs because the two disks must be parallel to each other.

Take the 1-inch wide aluminum strip and carefully cut it the exact length to fit around the perimeter of the north, or larger, disk from one side of the notch to the other. Drill a 1/8-inch hole every six inches along each edge of the strip and prepare the strip for countersinking the wood screws. Using 1-inch screws, screw one end of the strip to the edge of the disk next to the notch with one edge of the aluminum strip flush along the south face of the disk, the face towards the ribs. Work your way around the perimeter of the disk to the other end of the notch. Be sure the head of each screw is level with the surface of the aluminum strip. Use brass screws because they are soft and can be flattened somewhat if they protrude.

It is also appropriate to attach rubber stops to the ends of the

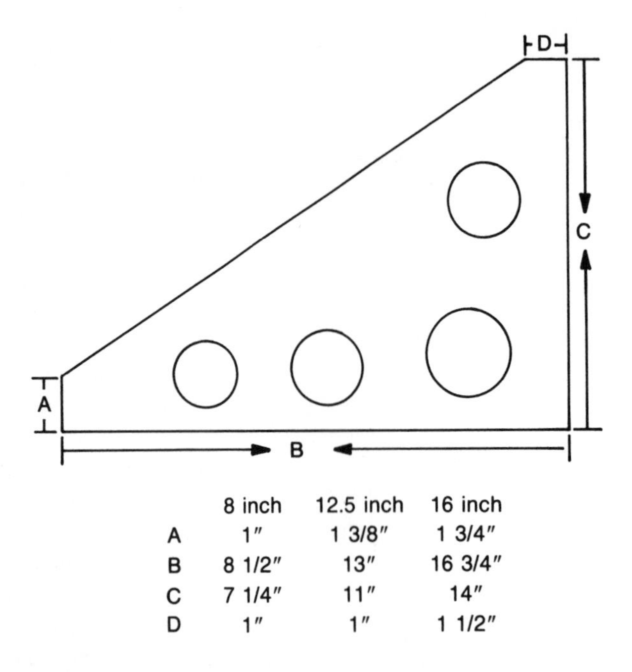

	8 inch	12.5 inch	16 inch
A	1″	1 3/8″	1 3/4″
B	8 1/2″	13″	16 3/4″
C	7 1/4″	11″	14″
D	1″	1″	1 1/2″

Fig. 3-8. Rib for polar disk.

aluminum strip to prevent the cone from rotating into the notch cut into the north disk. A suggested stop would be those used on sliding glass doors.

FINISHING THE BASE

Now that the polar cone is completed, you are ready to finish the base. First construct the brace that goes between the base arms. This brace, shown in Fig. 3-9, needs to be custom fitted to your base. The length of the brace should be the distance between the arms of the base measured at the south bearing. Cut the bottom part of the brace from a 2- × -4 and the vertical portion from a 2- × -4 for the 8-inch telescope, a 2- × -6 for the 12.5-inch and 16-inch telescopes. The height of the brace should equal the height of the arms of the base.

Next cut an arc in the vertical portion of the brace, making the center of the arc cut away approximately one-third of the width of the board. Attach the two pieces of wood together with screws and glue. Place the straight edge of the board with the arc to the face of the 2- × -4 even with one edge. Test fit the brace between the arms of the base. It must be exactly the same length as the distance between the arms at the south bearing.

Then cut two short pieces of 2- × -4 to support the steel wheels against the edge of the front disk of the polar cone. These supports should be the height of the vertical portion of the brace, because they fit into the corner between the arms of the base and the brace. Cut one end square and the other end at a compound angle. The angle across the thickness of the 2- x -4 should be equal to:

90 degrees minus your latitude

For example, our latitude is 46 degrees, so this angle is 44 degrees. This angle on the two pieces should be mirror images of each other.

The angle across the width of the 2- x -4 should follow the curve of the arc in the brace which should be approximately equal to the arc of the disk on the cone. You want the steel wheels to be nearly perpendicular to the edge of the disk.

To determine where to attach the brace to the base and where the support wheels for the edge of the polar cone should go, insert the EMT of the polar cone into the south bearing of the base. If the Teflon pieces are unattached, since you have not yet painted the base, carefully place the Teflon pieces in their proper places and tighten the south bearing. Center the cone between the arms of the base and support the polar cone such that there is no stress on the south end of the base and such that the length of EMT protruding from the south bearing is exactly perpendicular to the south end of the base. This placement of the polar cone is important because it must be at the proper angle to track accurately.

Without disturbing the position of the polar cone, place the brace between the arms of the base with the wheels held in place. Adjust the position of the brace until the wheels are in the center of the aluminum strip attached to the polar cone. See Fig. 3-10. This will place the rear edge of the polar cone approximately perpendicular to the rear edge of the horizontal 2- x -4 section of the brace. See Fig. 3-11 for more details. When you are sure of the positions of the brace and the wheels, mark them on the arms of the base and the 2- x -4 wheel supports. Remove the polar cone for

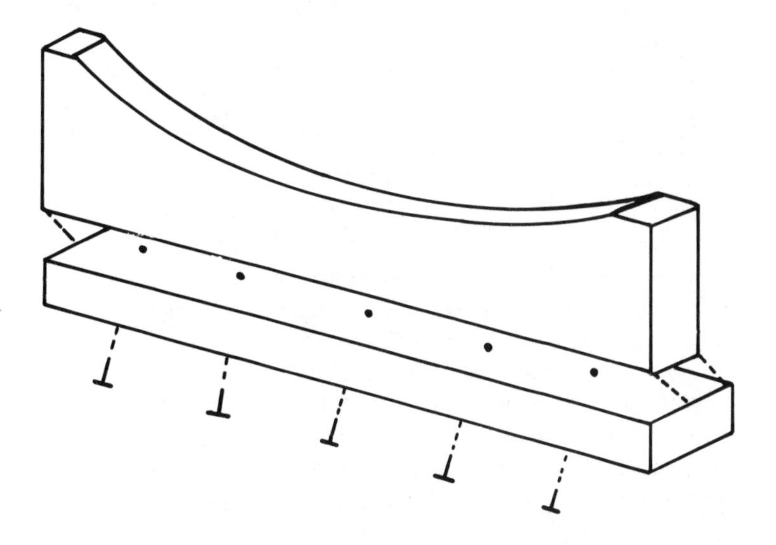

Fig. 3-9. Structure of base for brace.

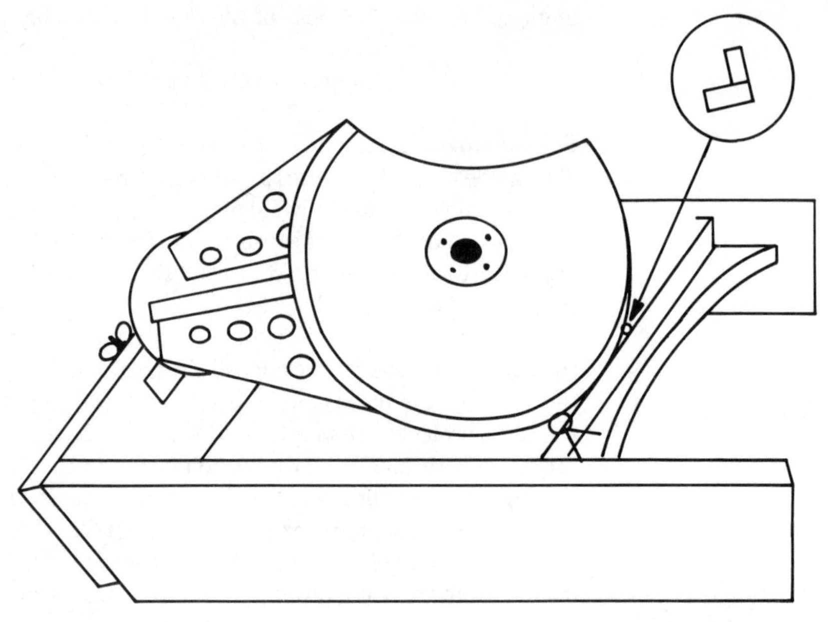

Fig. 3-10. Position of polar cone on base.

ease of handling and glue and screw the brace into place, making sure it is perpendicular to the arms of the base and even with the bottom sides of the arms. Then glue and screw the wheel supports into place.

To further increase the stability of your base, glue and screw some corner braces into the corners of the south end of the base. These braces can be diagonally ripped from scrap 2- × -2s or some other similarly sized boards.

Paint your base and attach the Teflon pieces and wheels. To increase the portability of the base, you can also attach two cabinet door handles. Your base is now finished, and you are ready to make the arms of the fork.

MAKING THE ARMS OF THE FORK

Figure 3-12 illustrates the shape and size of the arms of the fork. Cut four pieces of A-C plywood in this shape, sized appropriately for your telescope. Glue them in pairs with the C sides of the plywood together. When the glue is dry, fill the voids in the edge of the plywood and sand the edges smooth.

Each arm contains a bearing much like the bearing in the south end of the base. To make these bearings, cut a hole in each fork arm so that the edge of the hole is 1 1/2-inches from the top (side

Fig. 3-11. Position of wheels on large disk of polar cone.

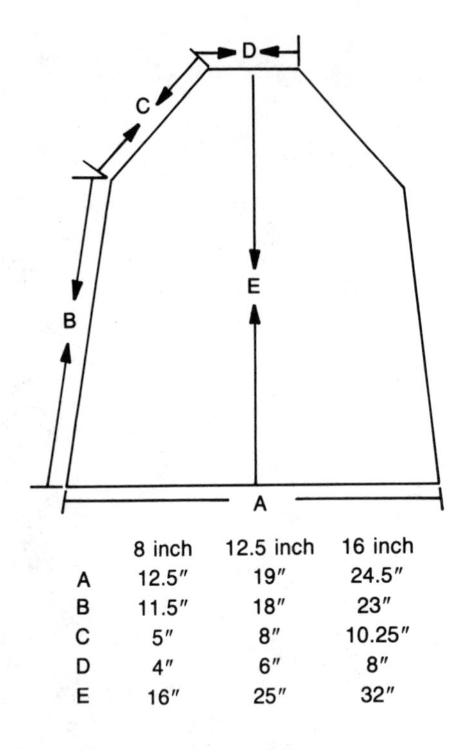

	8 inch	12.5 inch	16 inch
A	12.5″	19″	24.5″
B	11.5″	18″	23″
C	5″	8″	10.25″
D	4″	6″	8″
E	16″	25″	32″

Fig. 3-12. Fork arms.

"D" in Fig. 3-12) of each arm. This hole should be 1/2-inch larger than the outside diameter of the EMT used in your telescope and centered in the fork arm. See Fig. 3-13.

Next drill two 5/16-inch holes into the top of each arm. See Figure 3-14. These holes should be evenly spaced from the center at a distance 2 1/2-times the diameter of the hole you just cut. For example, we cut a 2-inch hole, so these 5/16-inch holes should be

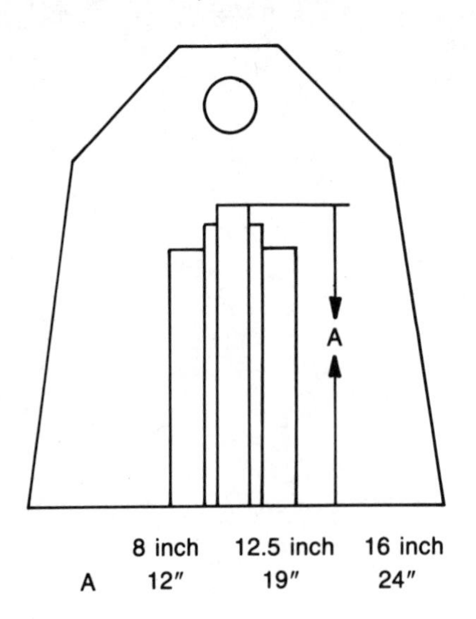

	8 inch	12.5 inch	16 inch
A	12″	19″	24″

Fig. 3-13. Reinforcing the fork arms.

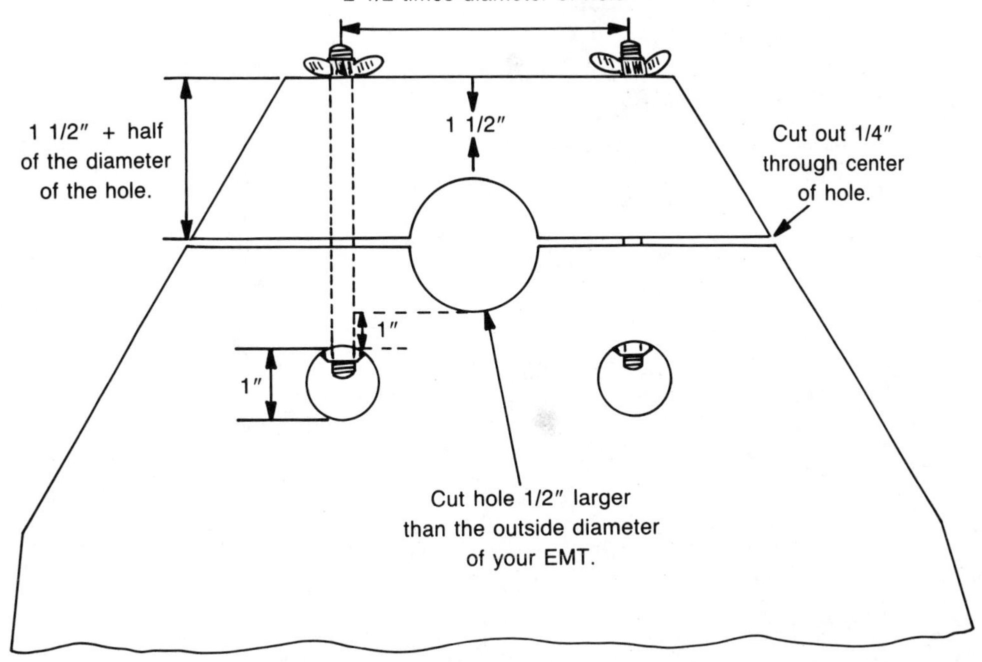

Fig. 3-14. Detail of declination bearing in fork arms.

drilled five inches apart, or 2 1/2-inches from the center of the top of the arm. Drill the holes at least 3 1/2-inches deep.

Draw lines on one face of each fork arm parallel to each of the holes you have just drilled. These lines should end at 2 1/2-inches plus the diameter of the first hole you cut below the top of the fork arms. Drill 1-inch diameter holes at the end of each line.

Draw a line through the middle of the first hole and parallel to the top of the fork arm. Now draw another line 1/8-inch on each side of this line. Cut along these lines so that the saw kerf is on the inside of this 1/4-inch wide strip. In other words, you want to cut off the top of the fork arm and at the same time remove 1/4-inch of its length.

The 5/16-inch holes you drilled should have gone past this cut. If so, finish drilling these holes into the 1-inch diameter holes drilled in the face of the fork. If not, carefully clamp the two pieces of each fork arm together. Be sure that the top piece is centered on the bottom piece. Drill the holes deeper. Then remove the clamp and finish drilling the holes into the 1-inch holes.

Cut a piece of Teflon plate 1 1/2-inches wide by 1 1/2-times the diameter of your EMT. Form and attach it to the lower portion of the fork arm as you did with the south bearing. Cut another piece of Teflon 1 1/2-inches by 3/4-times the diameter of the EMT. Form and attach it to the top of the fork arm.

Now cut four pieces of 5/16-18 threaded rod 1/2-inches longer

than the distance between the top of the fork arm and the top of the 1-inch hole in the face of the arm. For our 12.5-inch telescope, this was about six inches. Thread these rods through both sections of each fork. Place a small washer on the lower end of the rod and thread a nut on it. Place a section of EMT on the Teflon and put the top of the fork arm on this. Put another washer and a wing nut on the other end of the rod. Tighten the wing nuts and allow to stand for several days to help shape the Teflon.

The weakest link in the fork mount is the stability of the forks. To make them as rigid as possible, we built up the base of each fork arm to make it thicker. This was done by cutting some wedges from 2-×-4 and 2-×-3 lumber. First cut two lengths of 2-×-4 of length "A" as shown in Fig. 3-13. Glue these together and clamp until the glue is dry. Draw a line starting 1/8-inch from the corner on the unglued face of the wood to 1/8-inch from the opposite corner. Cut along this line. You will have two wedges with the glue joint running along the center of the diagonal cut.

Next cut 2 1-×-3s two inches shorter than the 2-×-4s you just cut. Diagonally cut these boards along the 3-inch faces just as you did the glued-up 2-×-4s. If you are making the 8-inch telescope, omit this step, while you should cut three identical pieces for the 16-inch telescope. Finally, cut a section of 2-×-4 four inches shorter than the first wedge, and diagonally cut it along the 2-inch face of the wood.

Sand all of these cuts smooth. Then glue them to the outside of the fork arm. Follow Fig. 3-13 for placement of the wedges of wood. Be sure that the cut sides of all the wedges of wood face away from the fork arm. Also make sure the thicker ends are level with the bottom of the fork arm. Once the glue has set, insert at least two 2-inch screws through the fork arm into each wedge of wood. Drill pilot holes through the fork arm and into the wedge. Countersink your screws. Figure 3-15 is a photo of the finished fork arm for our 12.5-inch telescope.

Paint or finish the fork arms at this time. The sides which will face each other should be a flat black. The outsides, which contain the reinforcement supports, can be any color you choose.

Set the fork arms aside while you begin the construction of the telescope tube. The fork arms will be mounted on the cone after you have made the central box of your telescope tube. You need the box to get exact measurements for the placement of the fork arms.

Fig. 3-15. Photograph of assembled fork arm.

Chapter 4

Constructing the Telescope Tube

THE NEXT STAGE IN CONSTRUCTING YOUR TELESCOPE IS TO begin making the tube. First, you will make the *trunnion box*, which is the center and pivot point of the telescope tube. After the trunnion box is completed, you then attach the fork arms to the polar cone. The final thing you do in this chapter is prepare the octagonal ends of the tube and the connecting lengths of EMT. The actual assembly of the tube comes after the optics have been installed.

MAKING THE TRUNNION BOX

The first step in making the telescope tube is to make the trunnion box. The assembled trunnion box is shown in Fig. 4-1 along with the major dimensions for the three sizes of telescopes. Begin by cutting four plywood rectangles with the dimensions "A" by "B" + 1-inch ("A" and "B" from Fig. 4-1). Now miter both of the "B" + 1-inch sides at a 45-degree angle, cutting off 1/2-inch from each edge in doing so.

Drill four 3/8-inch holes in each of the four sides of the trunnion box, two near the top and two near the bottom. To place the holes, measure in 1 1/2-inches from the top and bottom edges and 2 1/2-inches from the mitered edges. See side "E" of Fig. 4-1 for placement of these holes. Place a scrap of wood underneath the

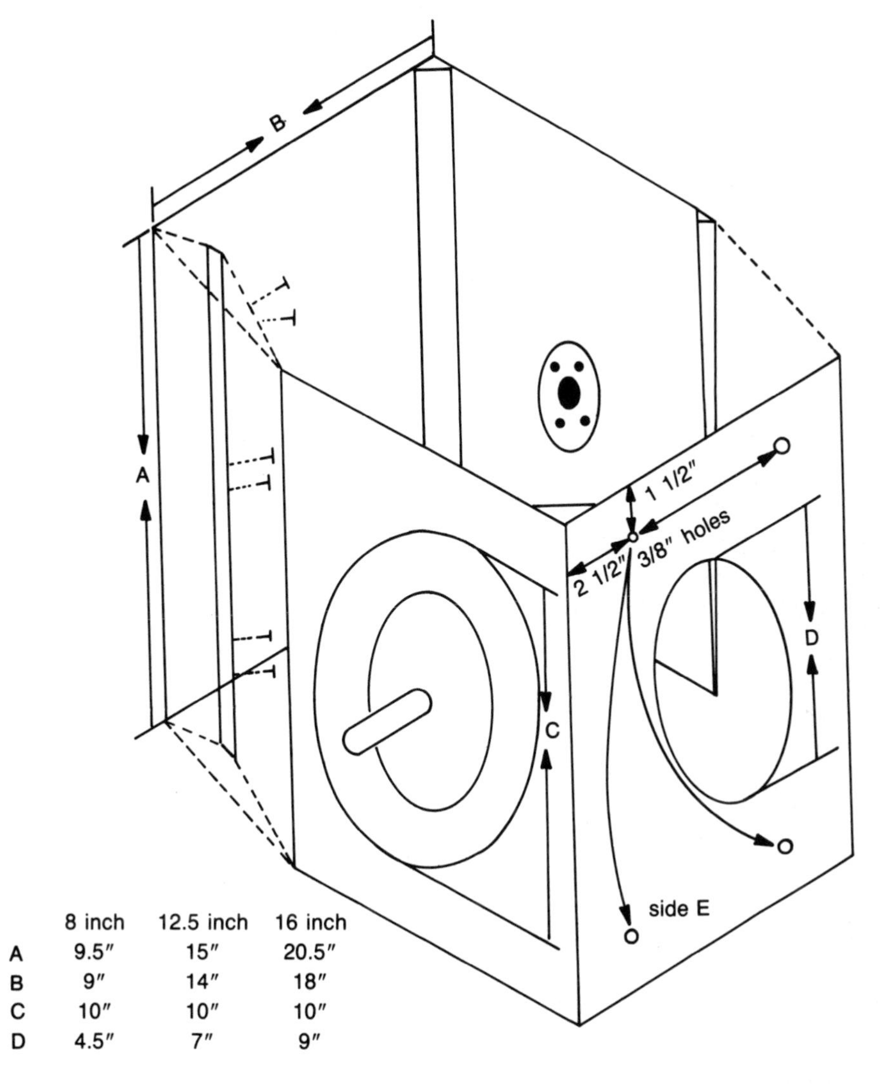

	8 inch	12.5 inch	16 inch
A	9.5″	15″	20.5″
B	9″	14″	18″
C	10″	10″	10″
D	4.5″	7″	9″

Fig. 4-1. Trunnion box (exploded view).

box side when drilling the holes to prevent the plywood from splintering. You will later use these holes to attach lengths of EMT as connectors between the trunnion box and the ends of the telescope tube.

Next cut three plywood disks, one with a diameter of "C" (Fig. 4-1) and two with diameters of six inches. Using contact cement, attach plastic laminate to one side of each of the two 6-inch disks, and weight them for two to three hours. When the contact cement is dry, cut a hole exactly in the center of each disk to fit the EMT used in your telescope. Also cut a hole the diameter of the EMT exactly in the center of two of the sides of the trunnion box.

Taking a short section of the EMT to use as a guide for assembly, insert it into the hole in one of the sides of the trunnion box, then into the hole in the 10-inch disk. The disk should be placed on what will be the outside of the trunnion box. Glue-and-screw the disk in place on the box side using 1 1/4-inch wood screws. Place the screws evenly around the disk 2 1/2-inch from the center of the disk. Countersink the screws so that the tops of the screw heads are level with the surface of the disk. Next glue one of the 6-inch disks to this assembly.

Repeat the above procedure with the second side of the trunnion box containing the hole in the center. Insert another short section of EMT into the hole in the side of the trunnion box and into the other 6-inch disk, then glue the disk to the outside of the box side.

While these are drying, cut a hole in each of the other two sides of the trunnion box. These holes should be diameter "D" from Fig. 4-1. There are two functions for these holes:

- Reduce the weight of the finished telescope. This function is more important for the larger telescopes.
- Allow you access to the inside of the trunnion box while constructing it. This function is more important in the smaller telescopes.

Using the template made for routing out an area for the pipe flange on the polar cone, rout out an area equal in depth to the thickness of the pipe flange on the inside of the two sides of the trunnion box with the disks. Make sure that this area is exactly centered over the hole already there.

Now make another template which is equal to the diameter of the router base plus the diameter of the shoulder of the pipe flange less half the diameter of your router bit. Center this one over the area just routed out and rout an area equal to the total depth of the pipe flange.

One problem about using a mix of plumbing and electrical fixtures is size incompatibility. The EMT cannot be threaded to fit the pipe flanges. To use both the plumbing and electrical fixtures, some work must be done to make them fit together. An adapter, such as was used in the polar cone, will not work in the trunnion box—it is too large for use. We used set screws which worked quite well.

To prepare the pipe flanges for use, drill three equally spaced 7/32-inch holes (use 5/32-inch for the 8-inch telescope) around the shoulder of each flange. See Fig. 4-2. Use fine-thread 1/4-inch (3/16-inch for the 8-inch telescope) hex head set screws and tap each hole. We found it necessary to remove a small amount of the thread from the inside of the pipe flange for the EMT to be inserted. Cut the EMT long enough to allow about 3 to 4-inches of the pipe to

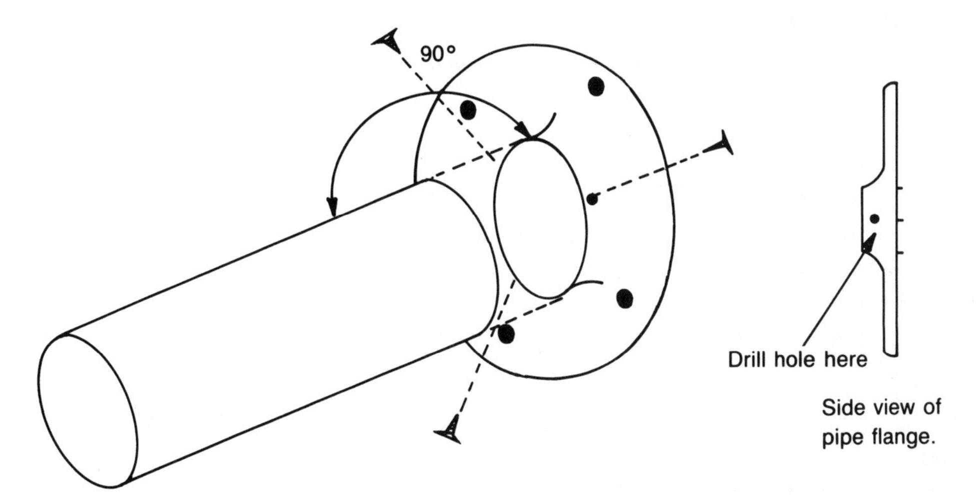

Fig. 4-2. Attaching EMT to pipe flange for the trunnion box.

extend beyond the disks attached to the sides of the trunnion box. Insert the set screws and tighten slightly onto a short piece of EMT, making sure that the EMT is perpendicular to the pipe flange.

Now attach the pipe flanges to the inside of the trunnion box in the area routed out for them with 1 1/4-inch pan head wood screws. Again make sure that the EMT is perpendicular to the sides of the box.

Preferably using a table saw or radial-arm saw, rip four pieces of 2- × -4 lumber of length "A" from Fig. 4-1 to make corner braces for the trunnion box. Set the blade at a 45-degree angle and rip off two corners of the board along the length of the lumber. When completed you will have a triangular shaped piece of wood with a base equal to the width of the 2- × -4 and a height at the center equal to the depth of the 2- × -4.

Drill six 1/8-inch holes through the wide side of the triangle. These holes are best made using a drill press. Place the wood on the drill press table with the narrow side of the triangle down. Drill three equally spaced holes about one-third of the way in from the edge. Reverse the wood and drill three more holes. Now using a 3/8-inch bit, redrill each hole only enough to countersink the screw heads. Insert 1 1/4-inch screws into each hole until the screws just protrude from the short side of the triangle.

Assemble the trunnion box by first attaching the corner braces to one corner of each side of the box. Glue-and-screw them so that the apex of the triangular brace is exactly aligned to the inside edge of the miter cut on the sides. Next assemble the four side pieces. The best order of assembly is to attach one of the sides with the large hole in it last, as you will need to reach through the hole to tighten the last screws into place.

During the assembly, be sure that the two pieces of EMT are exactly aligned. You can check this by sighting through one tube into the other. Adjust the box as necessary to assure alignment. Precise alignment is necessary because these pieces of EMT are the axles for rotation of the telescope tube about the bearings on the fork arms.

MOUNTING THE FORK ARMS

Now that the trunnion box is completed, you are ready to mount the fork arms onto the polar cone. The completion of the trunnion box was necessary as you need to know its exact size to get a close fit between it and the fork arms. If you are going to attach the computer controlled drive, then refer to the section on installing the declination drive in Chapter 6 before continuing here.

First measure the width of the trunnion box. Because accuracy is important here, carefully measure to the nearest 1/16-inch. In this measurement include the disks attached to the sides of the box, then add twice the thickness of the Teflon sheet you are using for bearing surfaces. The Teflon will be attached later. This measurement is the distance required for the trunnion box and Teflon-bearing surfaces to fit between the fork arms.

Support the polar cone so that the large disk of the cone is facing up. To locate the position for the fork arms, mark a line that extends through the center of the axle of the polar cone and is equal distance (distances one and two in Fig. 4-3) from the ends of the cut out arc. This is shown as line "A" in Fig. 4-3. Divide the number you measured as the width needed to mount the trunnion box by two. Now measure that distance from the center of the axle on either side of line "A". A center-finding straight ruler is helpful here. These distances are shown as points "B" in Fig. 4-3. Points "B" are where the mid-points of the face of the fork arms will be located.

To determine where to place the ends of the fork arms, divide the length of the bottom of the inside face of the fork arms (side "A" in Fig. 3-12) by two. Again using a center-finding straight ruler,

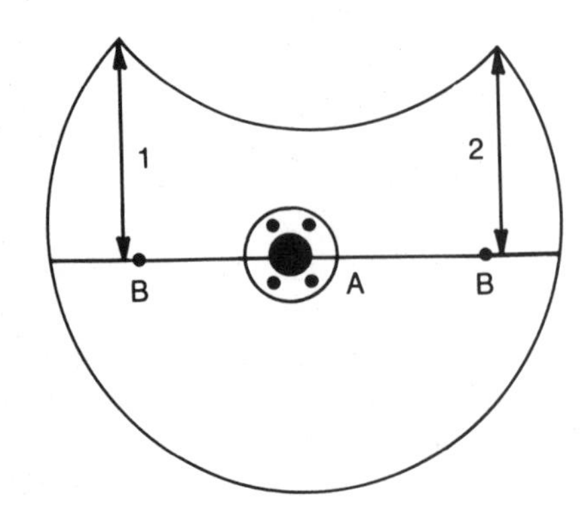

Fig. 4-3. Locating the fork arms on the polar cone.

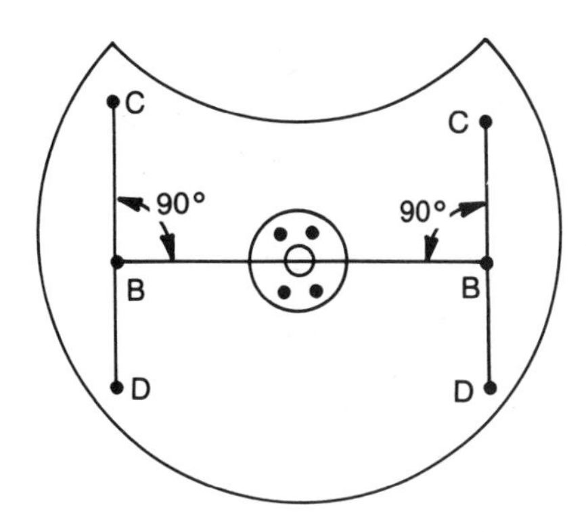

Fig. 4-4. Location of the fork arms on the polar cone.

place the center of the ruler on point "B," with the ruler perpendicular to line A. See Fig. 4-4. Mark a line, C—D, on both sides of the axle through points B, such that line segments C—B and B—D are both equal in length and perpendicular to line A. It is important that these lines not only be perpendicular to line A, but that they also be parallel to each other. To make sure that they are parallel, measure from point C to point C, point B to point B, and point D to point D. If the distance is the same between all three pairs of points, they are parallel. It is along these lines that the inside face of the fork arms will sit.

Place the fork arms on the face of the polar cone along the lines you have just drawn. Again, measure carefully the distance between the fork arms. If the distance is correct at both ends of the base of the arms, then outline the fork arms on the face of the polar cone.

Now drill six holes 13/16-inch deep in the bottom of each fork arm. These holes should be located approximately as shown in Fig. 4-5. Use 3/8-inch dowels (for the 8-inch telescope 1/4-inch will suffice) each 1 1/2-inches long. Using dowel centers, or some other locating device, mark the placement of the six holes in the face of the polar cone, making sure that the arm is exactly aligned within the area you marked above. Drill the holes in the polar cone.

Insert the dowels into the holes and test fit the fork arms. If the arms fit, place a small amount of glue in the holes in the face of the polar cone and insert the dowels. Using a wooden mallet, or a regular hammer with a block of wood, tap the dowels as far into the holes as they will go.

Coat the bottom of the fork arms and the holes there with glue. Press the fork arms down onto the dowels. Measure the distance between the tops of the fork arms. Make any necessary adjustments to ensure that the arms are the proper distance apart. Insert a piece of EMT into the Teflon-bearings on the arms and clamp it in place. Measure again to be sure you have the correct separation at the tops of the fork arms and set aside for the glue to dry.

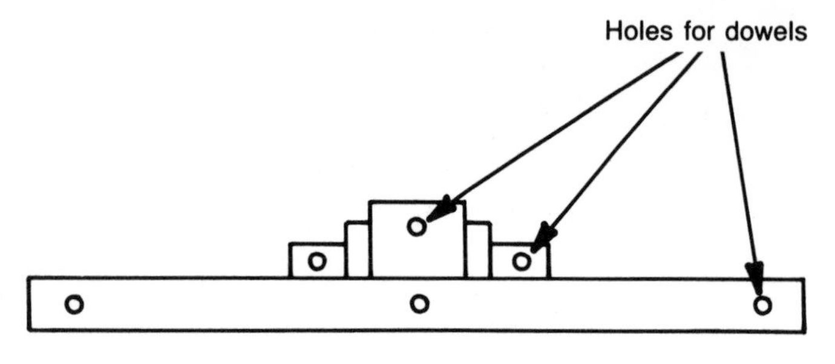

Fig. 4-5. Placement of the dowel holes in the fork arms.

MAKING THE ENDS OF THE TUBE

The trunnion box, which is the center portion of the telescope tube, is connected to the ends of the tube, which are plywood octagons, with lengths of EMT. To make the plywood octagons, you need to cut a number of identical parts. To ensure that all your cuts are made accurately, you should, whenever possible, make a jig to hold the wood in place, then carefully check the angles and wood sizes on a test cut before cutting all of the pieces. The following descriptions are for using a radial-arm saw. You can adapt the procedures for use with your saw if you do not have a radial-arm saw.

First, rip 6-inch and 3-inch wide strips of 48-inch long plywood from the A-B plywood. The number of strips of each width to be cut is listed in Fig. 4-6. Then cut 16 pieces 6-inches long and eight pieces 3-inches long from 1-×-2 lumber.

Adjust the saw blade to cut at a 22 1/2-degree angle. Then cut the 6-inch wide strips of plywood into 16 pieces that measure 6-inches wide by length A from Fig. 4-6. Flip the plywood over between cuts so that the bevel of the cut is in opposite directions, and endeavor to make the length of these cuts as close to identical as possible. Next cut eight pieces from the 3-inch wide strips by length A from Fig. 4-6. Again, flip the plywood over between cuts and make sure that these are the same length as the 6-inch wide pieces.

Next make the corner braces for joining the sides of the three octagons you are making. Without changing the angle of the saw blade, make a jig on the saw table like the one in the photograph in Fig. 4-7. Place two boards on the saw table so that they are parallel to the direction of travel of the saw blade and to each other. They should be positioned relative to the saw blade so that the blade will cut about 1/16-inch from the 2-inch side of the 1-×-2 and just far enough apart to allow free travel of a piece of 1-×-2 lumber between them. Once these two boards are correctly placed, clamp them down. Then clamp two more pieces of wood onto the outside of each parallel board, so that they are pressing against each side of the jig to prevent any spreading. Insert a backup board between the two parallel boards. This backup board will hold your 1- × -2s in place as you cut and will also prevent splintering.

When the jig is completed, check to make sure it is set up properly by making several trial cuts along the length of a scrap of 1- × -2. First cut along one edge of the 1- × -2, then flip it over and cut along the other edge to make a triangular shaped piece.

On the base of the 6-inch triangle, drill a 7/64-inch hole about one to 1 1/4 inches from each end on each side of the 6-inch braces. There will be a total of four holes in each piece. Now drill three holes in the 3-inch braces, two on one side about 3/4 to 1-inch from the ends and one in the center of the other side. Using a 5/16-inch drill, countersink each hole.

In one 6-inch width of plywood, make a hole of the proper diameter for your focuser. The hole should be centered with its edge 3/4-inch from one of the unmitered sides of the wood. For the Lumicon focuser, which we used, this was a 2 1/8-inch diameter hole. Also drill holes for the bolts needed to mount the focuser. Use tee nuts on the inside of the piece for these mounting bolts. Next drill the 3/8-inch holes in four of the 3-inch wide plywood pieces and four of the 6-inch wide plywood pieces that will later be used to connect the lengths of EMT. Each of these pieces of plywood should

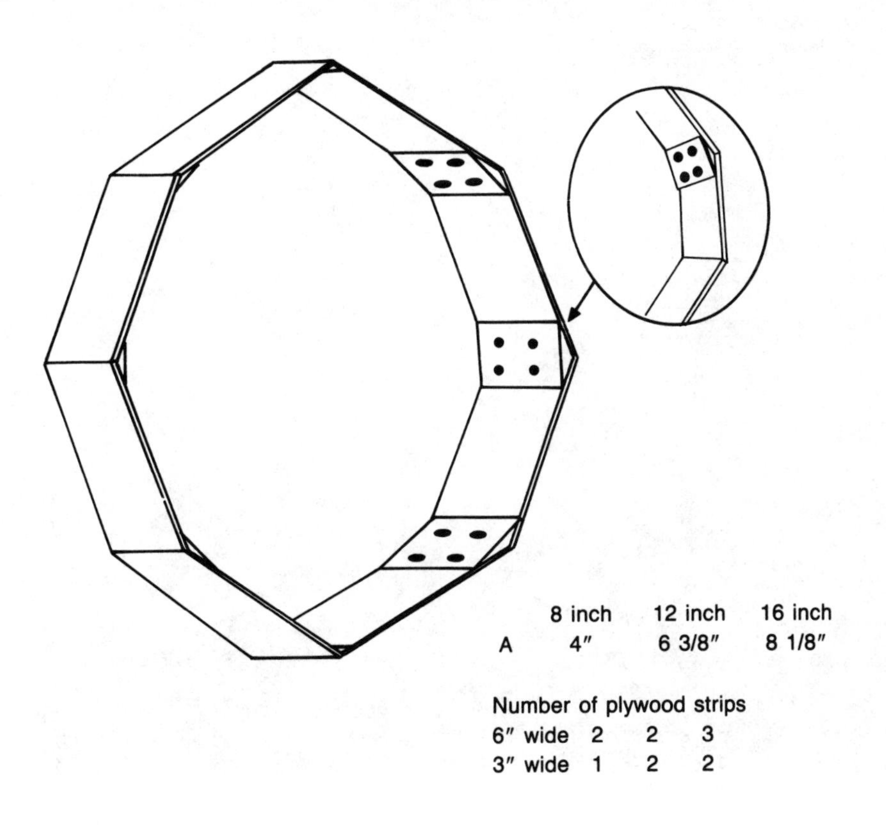

Fig. 4-6. Octagon construction.

Fig. 4-7. Jig for cutting the triangular corner braces.

have two holes drilled 1 1/2-inches from each mitered side and 1 1/2-inches from an unmitered edge. During assembly, be sure that these holes are toward the same edge of their respective octagons. The piece with the focuser hole should not be in the same octagon as the pieces with the 3/8-inch holes.

Assemble the octagons by first making a jig like the one shown in the photograph in Fig. 4-8. This jig is made from a scrap of 1-×-2 and a scrap of 2-×-4 each at least eight inches long. Clamp the 1-×-2 to the table. Then to find the position for the 2-×-4, lay one of the sides of the octagon on the table with a mitered side pressing against the 1-×-2. Lean another piece of the octagon against the 2-×-4 and adjust the position of the 2-×-4 until the joint between the two plywood pieces is together the way it should be. At that point, clamp the 2-×-4 to the table top.

With the jig ready, you are ready to construct the octagons. Place one of the plywood sections into the jig. Coat a mitered edge of another plywood section with glue and place it against a mitered edge of the piece in the jig. Adjust the two pieces until they are

Fig. 4-8. First jig needed to assemble the octagons.

lined up properly. Insert the screws into a corner brace and mark the mitered pieces with the locations for the screw holes and make starter holes with an awl or screw starter. Then coat the joining surfaces of the brace with glue, put it in place, and tighten the screws. Continue attaching sections of plywood together until they are all joined in pairs. Be careful as you work—if the pieces slip during assembly, the joints will not be smooth nor will the octagon be symmetrical.

Next assemble these sections together to make halves of octagons. To make your work easier, construct a jig like that in the photograph in Fig. 4-9. The distance between the 2- × -4 scraps is just far enough to hold the two pieces together. Assemble the joints by following the same procedure as with the first pairs.

Finally, assemble these halves of the octagons into complete octagons. We found that it was easiest to fix them between two stops clamped to the table. The stops press the two halves together from the center of the assembled sides. Glue-and-screw the remaining braces into the two joints.

Fig. 4-9. Second jig needed to assemble the octagons.

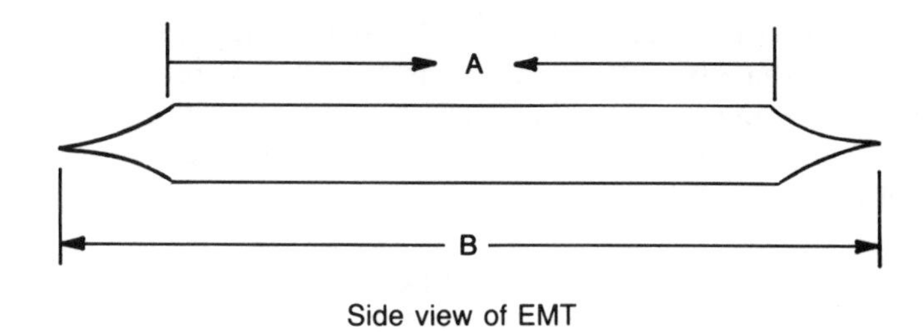

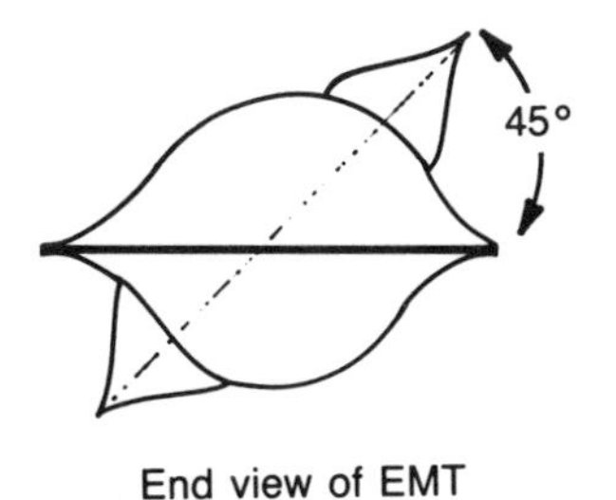

	8 inch	12.5 inch	16 inch
A	5″	8″	10 1/4″
B	9″	12″	14″

Fig. 4-10. The EMT for the tube braces.

A useful design feature of this telescope is the ability to turn the octagon which contains the focuser to any of the eight positions possible on the end octagon. This feature will help prevent some of the awkward positions astronomers sometimes find themselves in when observing. Drill two 3/8-inch holes 1-inch deep 1 1/2-inches from the mitered edges in the ends of a set of opposite sides of the 3-inch wide octagon. Glue four 3/8- × -2 1/2-inch dowels into these holes. When mounting this octagon onto the EMT, the edge with the dowels should face away from the trunnion box. Then drill two matching 3/8-inch holes 1 1/2-inches deep into each edge piece of the octagon containing the focuser. Make sure these holes are on the side of the octagon which will face toward the trunnion box. The octagon containing the focuser will now fit onto the dowels in the 3-inch wide octagon.

MAKING THE TUBING BRACES

Cut eight pieces of EMT of length B from Fig. 4-10. Flatten two inches of both ends on each piece. First flatten one end of all the pieces of EMT. The other end should be flattened at a 45-degree angle with respect to the first end, but the angle should face 45-degrees in one direction for four of the pieces and the other direction for the other four pieces. See end view of EMT in Fig. 4-11. Drill a 3/8-inch hole in the flattened sections 1/2 inches from the ends. The flattened ends will need to be bent slightly because the trunnion box is somewhat smaller than the octagonal mirror cell.

Cut another set of eight pieces of EMT. The method for calculating the length of these pieces is rather complicated, but it is important to follow these directions carefully. First, find the focal length of your mirror. This is easy, if you have the diameter and the focal ratio—simply multiply the diameter of the mirror by its focal ratio. For example, the focal length of a 12.5-inch f/5 mirror is 62.5 inches. Hold on to that number while you total the following lengths:

- The length of the tubing connecting the mirror holder and the trunnion box. This is A from Fig. 4-11.
- The length of the trunnion box. This is side A from Fig. 4-1.
- The width of the narrower octagon (about three inches).
- Half the distance between the opposite sides of the octagon holding the focuser. Measure from the approximate center of one flat plywood section to the approximate center of the plywood section opposite it and divide by two.
- The distance between the top of the mirror and the top of the mirror-cell holder. This distance is approximately one inch. Refer to Chapter 5.
- The radius of the lower edge of the focuser holder.

When these numbers are added, subtract half of the length of the fully extended focuser. Now subtract the resultant number from the number you calculated as the focal length of your mirror. For our final value we obtain 25 1/2 inches. To this value, add four inches to allow for the flattened sections on each end of the tubes. For our example telescope, the length calculated for cutting the EMT is 29 1/2 inches. Carefully measure for your telescope. Your numbers may be slightly different.

When you have measured and cut the tubing for the top section of the telescope tube, flatten two inches on each end. Make sure that—like the shorter braces—the flattened sections are at a 45-degree angle to each other with four angled one way and four angled the other way. Drill a 3/8-inch hole 1/2 inches from the flattened ends of the EMT.

Final assembly of these items will continue after the construction and installation of the optical assemblies, described in the next chapter.

Chapter 5

Installing and Collimating the Optics

YOU ARE NOW READY TO BEGIN THE LAST STAGES OF BUILDing your telescope. When you have completed the steps in this chapter, you will have a completely functional telescope. In this chapter you will construct the primary mirror-cell, the secondary mirror holder and assemble all the various parts of the telescope.

PAINTING YOUR TELESCOPE

If you have not been painting the various parts of your telescope at the points we suggested during construction, you should consider doing so now. A thorough coat of paint is essential if you live in a damp climate, and recommended even if you live in a drier climate. Because wood will swell when it absorbs moisture and shrink when it dries, you will want to make every effort possible to prevent this change in size in order to protect the optical alignment of your telescope.

Before painting, seal all wood surfaces with a latex primer. Paint the parts of the telescope not visible to the mirror with a good grade of exterior latex paint. Use any color you like, but avoid high-gloss paint. We used a light gray. Paint the surfaces of the telescope visible to the mirror—the face of the polar cone, the insides of the fork arms, the inside of the trunnion box, the insides of the octagons, the primary mirror-cell, and the secondary mirror holder—a flat black.

Be sure to paint before installing the optics. Paint splatters can ruin the optical surfaces.

BUILDING THE PRIMARY MIRROR-CELL

The primary mirror-cell is the part of your telescope which holds the larger of the two mirrors. Although primary mirror-cells are available from a variety of sources and are often inexpensive, because of the simplicity of their construction, we elected to make ours. It is important that the primary mirror be solidly supported, yet the supports should touch it as little as possible. To accomplish this, we used a nine-point flotation mirror-cell. If you decide to purchase a mirror-cell, skip this section.

For all plywood pieces in this section, use A-B grade plywood.

Construct the base for the mirror-cell by cutting a plywood disk with a diameter 1/2-inch larger than the diameter of your mirror.To attach the mirror clips, make three marks equally spaced around the edge of the disk, then drill a hole 1-inch deep into the edge of the disk at each of these marks. Because there are a number of types of threaded inserts, use the appropriate drill bit for the insert you're working with. Ours required a 7/32-inch hole. Insert a 10-32 threaded insert into each of these holes. Be careful here—many threaded inserts have a slot which looks as if it can be driven with a screwdriver, but unfortunately, using a screwdriver will often cause the insert to break. Instead, thread a nut onto a 10-32 screw about 1/4 -inch, then thread the screw into the insert. Tighten the nut against the insert and use the slot on the screw to drive the insert.

Draw a line from the center of the disk to each of the inserts in the edge of the disk. From the center, measure distance A from Fig. 5-1 along these lines and, except for the 8-inch telescope, drill a 3/16-inch hole at each of these points. Draw another three lines from the edge of the disk to the center, halfway between the inserts, or if you are making an 8-inch telescope, halfway between the marks. On these lines, measure distance B from the center of the disk, and at those points drill 1/4-inch holes. If you are making an 8-inch telescope, drill these holes. Countersink the 1/4-inch holes by drilling them again 3/8-inch deep by 3/4-inch in diameter.

For the 8-inch telescope, after you drill and countersink the 1/4-inch holes, cut three 1/2-inch square pieces of 3/4-inch plywood and glue them to the disk at the points marked by length A in the previous paragraph. Insert a 1 1/4-inch screw from the other side of the disk into each plywood block. Now cut three 3/8-inch square pieces of Teflon and, using a 3/4-inch brad, nail it to the 1/2-inch plywood blocks. Be sure to use a nail set to drive the head of the

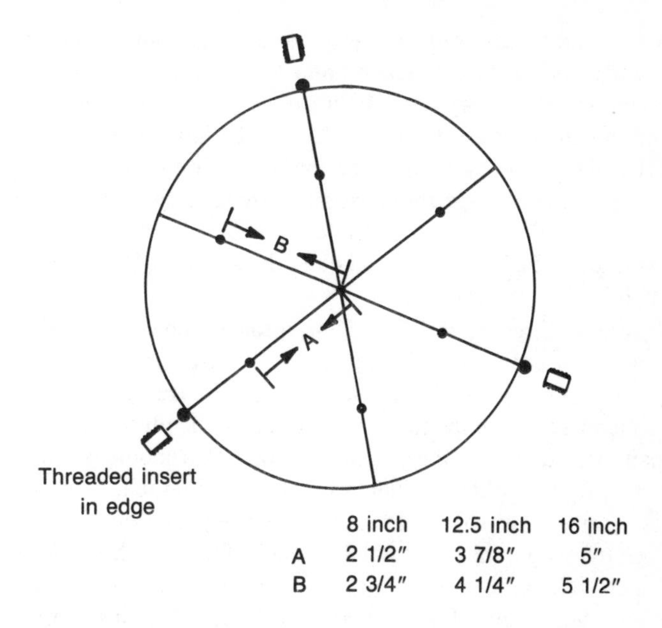

	8 inch	12.5 inch	16 inch
A	2 1/2″	3 7/8″	5″
B	2 3/4″	4 1/4″	5 1/2″

Fig. 5-1. Placement of holes in the mirror base disk.

brad at least 1/16-inch below the surface of the Teflon. Skip the next three paragraphs.

For the 12.5-inch and 16-inch telescopes, cut three plywood triangles of the dimensions shown in Fig. 5-2 and nine 1/4-inch square pieces of Teflon plate. Using 3/4-inch brads, nail the Teflon squares to the plywood triangles at the distances C and D shown in Fig. 5-2.

Now cut three pieces of Teflon plate 1/2-inch square. In the center of each Teflon square, drill a 11/32-inch hole just deep enough for the 10-32 cap nuts to fit almost to the shoulder of the nut. Using a propane torch, heat each of the holes and insert a spare cap nut into the hole while the Teflon is hot. Tap lightly with a hammer and allow to cool. Check to see if the hole is formed to fit the nut so that the Teflon and cap nut can act as a ball and socket joint. Repeat as necessary until the Teflon and cap nut fit properly.

Insert 1-inch long 10-32 screws into each of the 3/16-inch holes

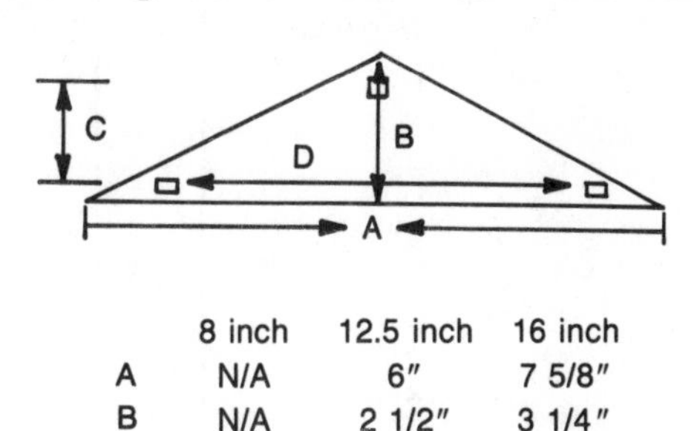

	8 inch	12.5 inch	16 inch
A	N/A	6″	7 5/8″
B	N/A	2 1/2″	3 1/4″
C	N/A	2″	2 1/2″
D	N/A	5″	6 1/2″

Fig. 5-2. The triangles for the mirror-cell.

in the disk and thread into the cap nuts. The cap nuts should be tight against the plywood. Place one of the 1/2-inch squares of Teflon on one of the cap nuts. Balance a plywood triangle, with the Teflon pads facing up, on the 1/2-inch Teflon square. Adjust the plywood triangle until it balances on the Teflon square and cap nut. Then carefully mark the location of the Teflon on the triangle. Attach the Teflon to the triangle with a 3/4-inch brad. Using a nail set, drive the nail 1/16-inch below the surface of the Teflon. Repeat for all three triangles.

Next make three mirror clips. First, measure the vertical height from the center of the threaded inserts in the edge of the plywood disk to the top of the small Teflon pieces on the triangles, or, for the 8-inch telescope, to the top of the Teflon on the plywood blocks attached to the disk. For the 8-inch telescope, this should be about 1 3/8-inch, for the others it should be about 1 7/8-inch. Next add to this number the thickness of the primary mirror. For the 12.5-inch telescope, this will total 3 1/8-inch. Then cut three pieces of 3/4-inch wide flat iron 1 1/4-inches longer than the total arrived at for your telescope. Make a 90-degree bend 3/4-inch from one end. See Fig. 5-3. Drill a 1/4-inch hole in the longer arm of the metal strip at a point half the thickness of your mirror from the bend.

Next make a slot 1/4-inch wide by 3/4-inch long positioned so that, with the short arm of the metal strip resting on the surface of the mirror, the threaded insert will come about midway in the slot. One way to make this slot is to drill three 1/4-inch holes as close to each other as possible. Then use a file to smooth the inside of the slot. Also file off all metal burrs found on the piece. Glue a piece of felt to the inside of the short arm of the mirror clip.

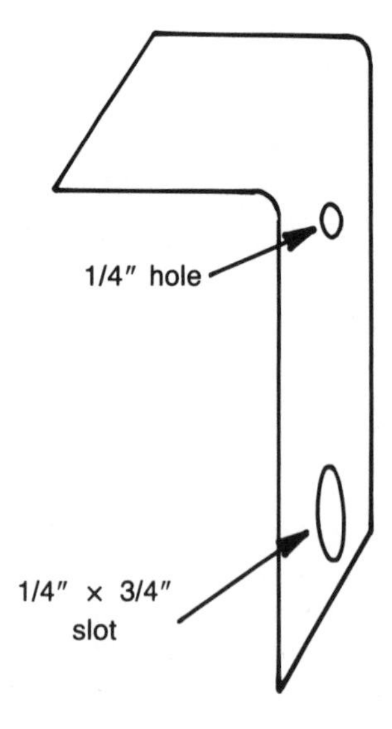

Fig. 5-3. A mirror clip.

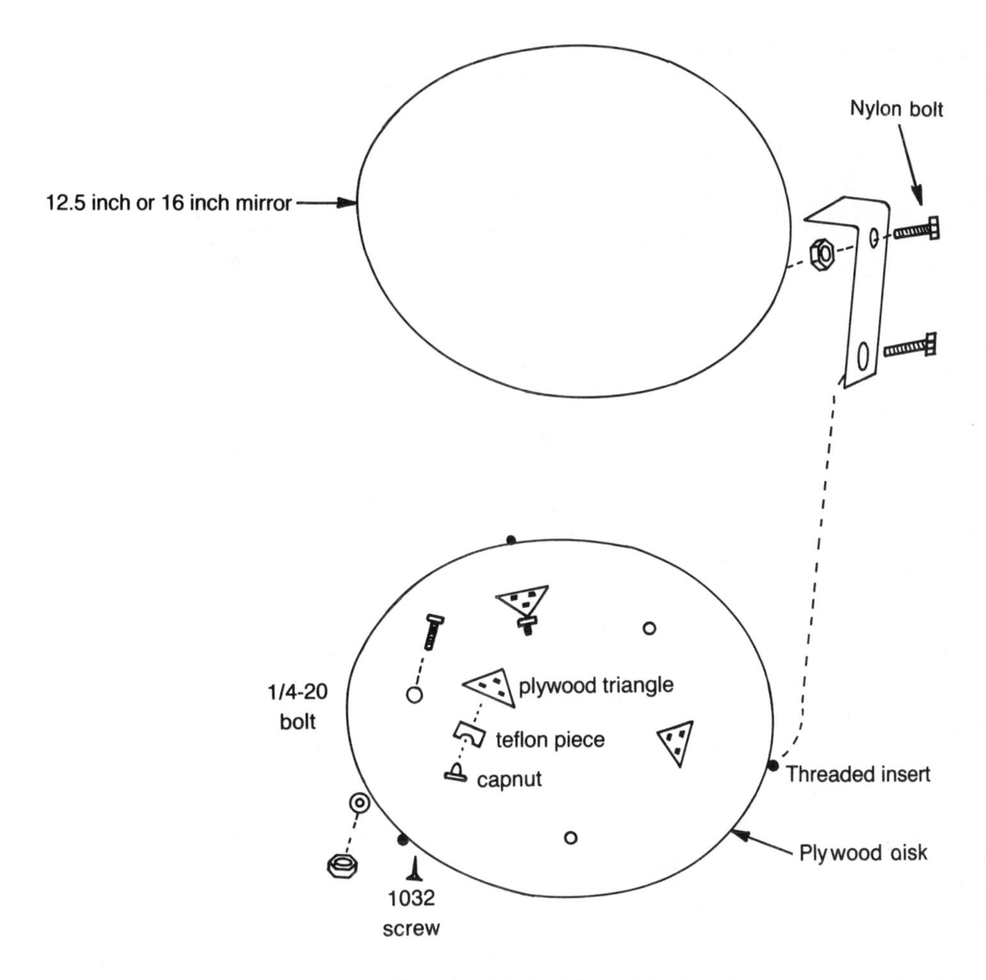

Fig. 5-4. Exploded view of the mirror-cell for the 12.5- and 16-inch telescopes.

Insert a 1/2-inch long 1/4-20 nylon screw into the 1/4-inch hole in the mirror clip. Our screws were called *license plate screws* at the hardware store. Thread, but do not tighten a nylon nut onto it.

Paint the disk and associated parts flat black. After placing a washer into the 3/4-inch holes, insert a 3-inch 1/4-20 bolt into the 1/4-inch holes. The head of these bolts should be toward the mirror. See Fig. 5-4 (Fig. 5-5 for the 8-inch telescope). Put a washer and nut onto the bolt and tighten so that the bolt is firmly in position. Place another nut onto the bolt. Tighten to within 1-inch of the surface of the plywood. As you place these nuts on each of the bolts, tighten them so they are all approximately the same distance from the plywood.

Arrange the plywood triangles on the cap nuts as shown in Fig. 5-6 and carefully lay your mirror on top of them. Center the mirror over the plywood disk. Gently place the mirror clips in their proper location and tighten 1-inch long 10-32 screws into the threaded inserts in the edge of the disk. The mirror needs to be held snugly

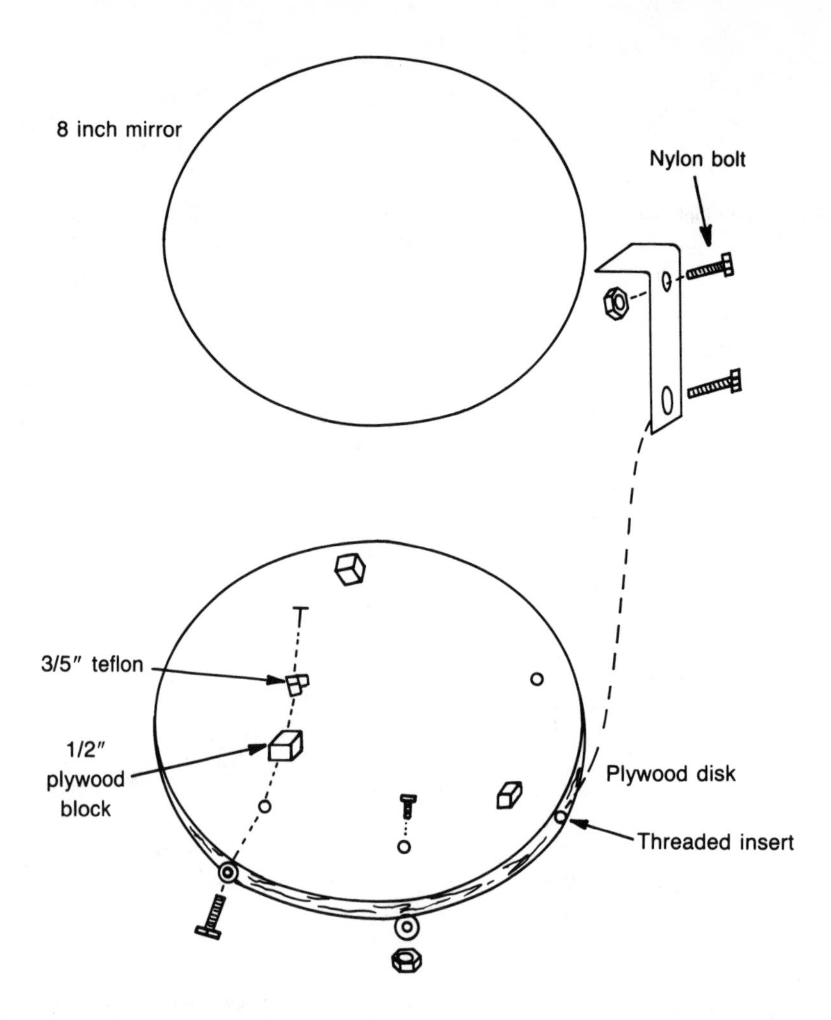

Fig. 5-5. Exploded view of the mirror-cell for the 8-inch telescope.

in place, without placing any strain on the mirror. Tighten the nuts on the nylon screws until they press tightly against the edge of the mirror. Set this assembly aside away from the dust of your workshop.

Next, make the backup plate of the mirror-cell. The backup plate is the portion of the mirror-cell which attaches the mirror and its base to the octagon. Cut a plywood square of dimension A in Fig. 5-7. There are two ways you can shape the plywood square. First, you could make a simple octagon and skip the rest of this section. An octagon is acceptable if your mirror will seldom experience any large temperature changes. Second, you could make a shape like the design of the piece shown in Fig. 5-7. This shape is more desirable because it will allow better air circulation than the simple octagon, thus making it less susceptible to temperature changes.

If you choose to make a piece like Fig. 5-7, follow these direc-

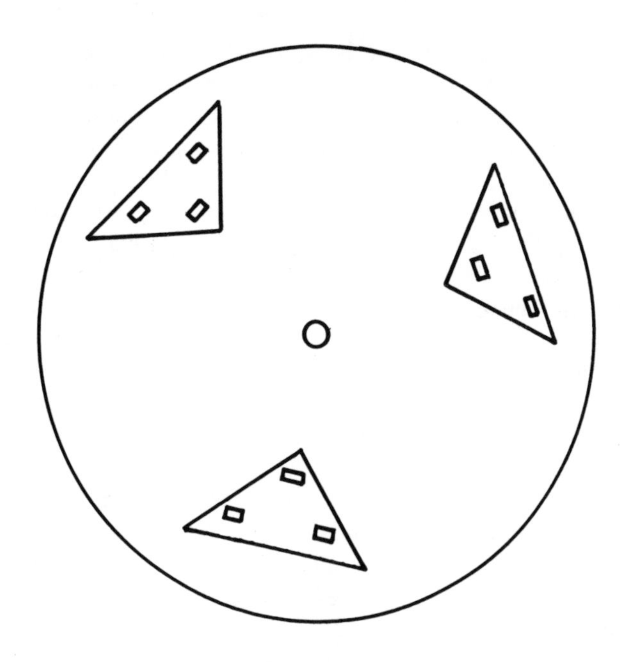

Fig. 5-6. Arrangement of the triangles in the nine-point cell.

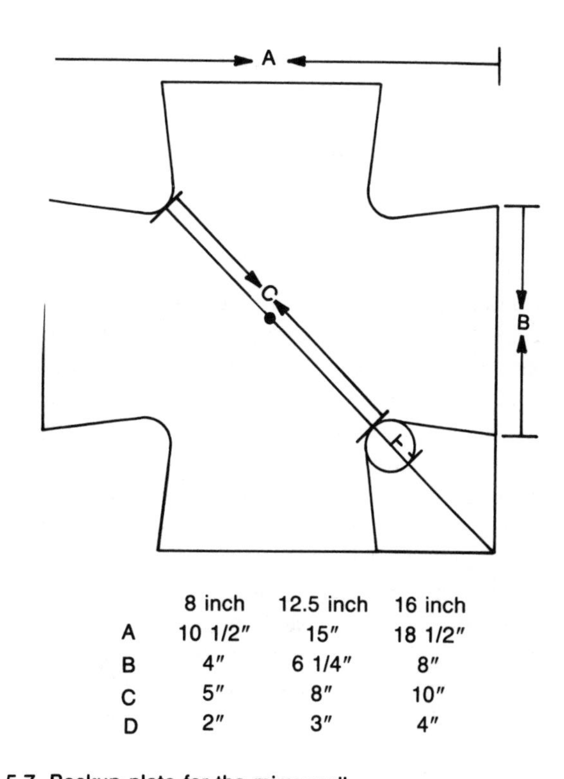

	8 inch	12.5 inch	16 inch
A	10 1/2″	15″	18 1/2″
B	4″	6 1/4″	8″
C	5″	8″	10″
D	2″	3″	4″

Fig. 5-7. Backup plate for the mirror-cell.

tions to cut the corners out of the square. First, find the center of each side of the plywood square, and measure a distance equal to half of length B on either side of those center points. Next draw diagonal lines from corner-to-corner of the square. From the center of the square, the point where the two diagonal lines cross, make a mark at a point half of distance C from the center towards each corner along the diagonal lines. A square cut corner would cut away too much wood, so round the corners. To do this use a compass, and draw a circle of radius D in each corner so that the edge of each circle is at the marks at the ends of line C and the center of each circle is on the diagonal line between the end of line C and the corner of the square. Now draw lines from the marks that indicate the ends of line B at the edge of the square to the edge of the circle. Cut along that line to remove the corner.

Next, drill a set of three 5/16-inch holes centered on the center of the backup plate and matching the 1/4-inch holes in the plywood disk backing up the mirror. Also drill two holes in edges B (Fig. 5-7) 1 inch from each corner to fit threaded inserts for 1/4-20 bolts. Paint the backup plate flat black.

When the paint is dry, insert the 1/4-inch bolts projecting from the mirror-cell into the holes in the backup plate. Thread two nuts onto each bolt and tighten. Figure 5-8 is a top view photograph of the completed assembly. Figure 5-9 is a side view photograph of the completed assembly. The wooden parts were left unpainted to provide better contrast for the pictures. Also you will notice in Fig. 5-8 that there are holes in the top and sides of the mirror clips that you were not asked to make. Ignore them, please, they came along with the metal strips we bought to make the clips.

MAKING THE SECONDARY MIRROR HOLDER

The secondary mirror is necessary to converge the image from the primary mirror and reflect that image out the eyepiece. Secondary mirror holders and their associated spiders are usually inexpensive and available from various sources, but because of the simplicity of construction, we have elected to make ours. If you decide to purchase one, skip this section.

Although the instructions for the secondary mirror holder look complicated, in reality they are not. Study the drawings and photographs before you begin, follow the directions carefully, and you should have no problems constructing your secondary mirror holder.

For all plywood pieces in this section, use A-B grade plywood.

First, make the mirror pad. Glue two 6-inch lengths of 2- × -4 scraps together to form a 4- × -4 square. For the 16-inch telescope, use three 6-inch lengths of 2- × -6 scraps. The wood should be clear grained and not warped. Clamp the pieces tightly while the glue is drying.

Fig. 5-8. Top view photograph of the completed mirror-cell.

Fig. 5-9. Side view photograph of the completed mirror-cell.

When the glue is dry, trim the wood to form a square such that the width of each side is slightly less than the minor axis of the secondary mirror. See Fig. 5-10 to visualize the axes. As you cut, keep in mind that the glue joint is to be centered lengthwise in the pad. We are making a 12.5-inch telescope, with a secondary mirror having a minor axis of 3.1 inches, so we cut the wood 3-inch square. For an 8-inch telescope, the block should measure 2 1/2-inch square, and for a 16-inch telescope, the block should measure 4 1/8-inch square. Next, trim the corners of the block with a 45-degree cut to make an octagonal cross section. See Fig. 5-11. Then cut one end of the wooden octagon to make sure that it is smooth and that the bottom is perpendicular to the sides.

Drill a 3/8-inch hole exactly in the center of the smoothed end of the octagon at least 3 1/2-inches deep. Then cut the unsmoothed end of the octagon at a 45-degree angle. The block should be positioned so that the glue joint is parallel to the saw table for a table

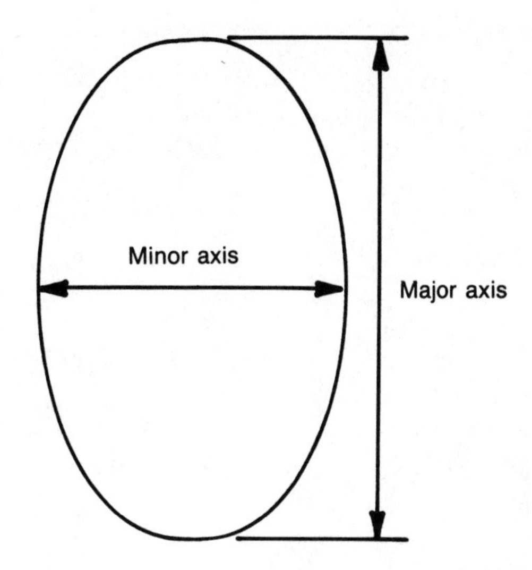

Fig. 5-10. Secondary mirror.

saw or radial-arm saw. Make the height of the short side of the cut 3/4-inch from the end of the octagon containing the 3/8-inch hole. After you make the angled cut, the height of the long side will be about four inches. The 3/8-inch hole which you drilled from the other side should be visible on the angle cut. Using a 1-inch hole saw or drill bit, cut a hole centered in the 3/8-inch hole in the angled side to countersink the bolt you will use to hold the pad in place. Drill the hole until the entire bit is just below the surface of the wood. If you used a hole saw, chisel out the center of the hole. Fig. 5-12 is a front view of the completed secondary mirror holder. This view gives you a good idea as to what the mirror pad should look like.

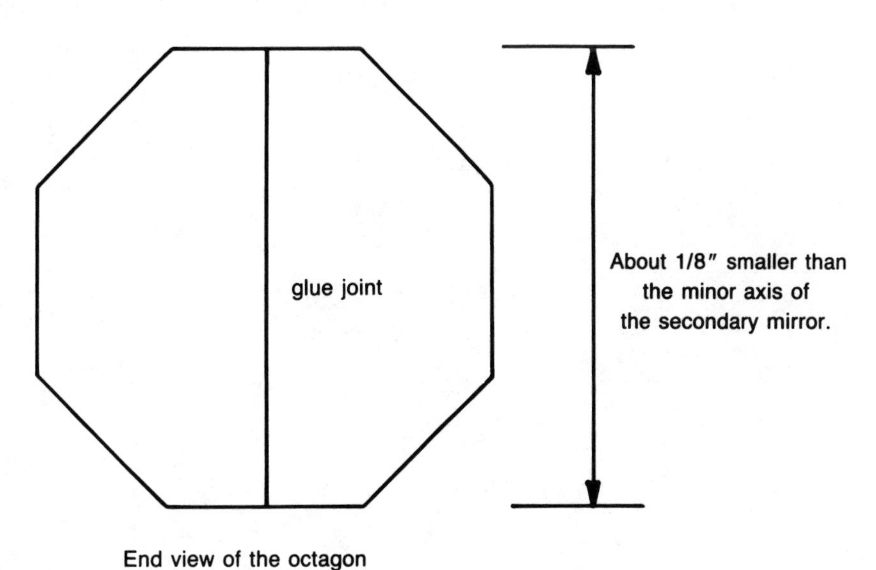

Fig. 5-11. Octagon for secondary mirror holder.

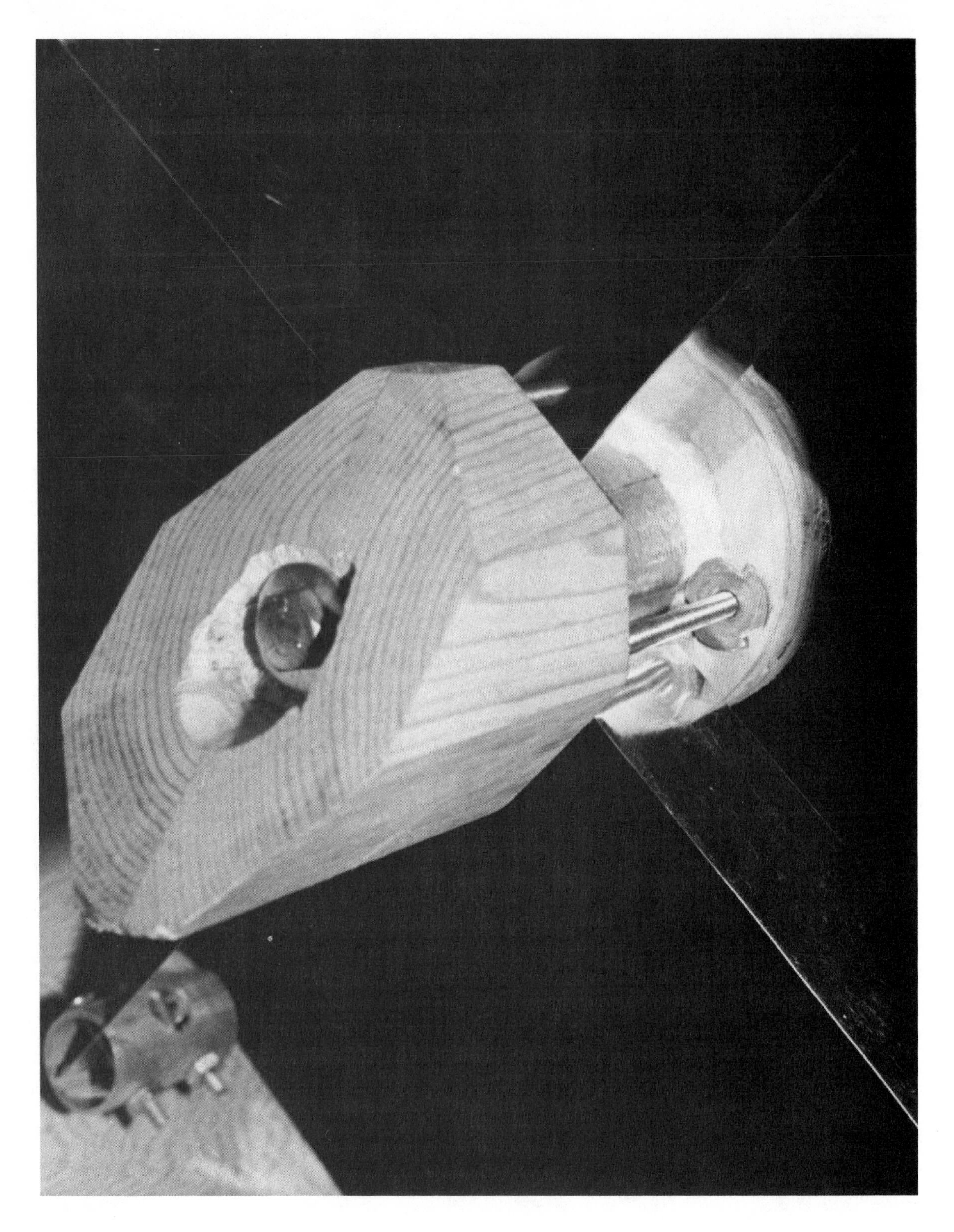

Fig. 5-12. Front view photograph of the completed secondary mirror holder.

Cut a piece of metal to just fit the flat end of the octagon. Using contact cement, glue the metal to the octagon. Then using 1/2-inch #6 pan head wood screws, screw the center of each of the eight edges of the metal to the wood. Set this assembly aside. Figure 5-13 is a rear view of the completed secondary mirror holder. This view helps you see how to attach the metal plate.

Now you are ready to make the spider portion of the secondary mirror holder. Cut a 1 7/8-inch section of the larger EMT, making sure that the ends are parallel and smooth. Use a hacksaw and cut four slots at right angles to one another from one end to within 3/8-inch of the other end. Clean the burrs from the cuts you just made.

Take the 1 1/2-inch sheet metal strips and thread them through adjacent slots in the EMT to make the spider vanes. See Fig. 5-14 for details. Cut these vanes length A measuring from the outside of the EMT.

Fig. 5-13. Rear view photograph of the completed secondary mirror holder.

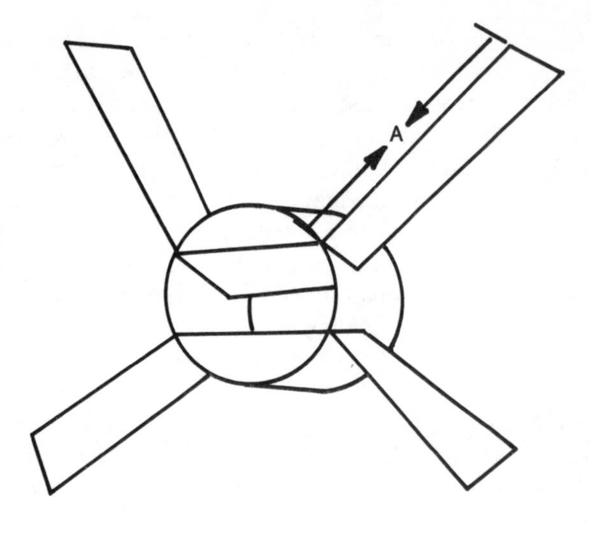

	8 inch	12.5 inch	16 inch
A	3 5/8″	5 3/4″	7 3/8″

Fig. 5-14. Threading design for the vanes of the spider.

Using the 3/4-inch EMT, cut four pieces 1 1/2-inches long. Using a 1/4-inch drill bit, drill completely through both sides of the EMT in the center of the length of each piece. Now enlarge one of these holes in each piece to 1/2-inch. Next drill two sets of 3/16-inch holes completely through the EMT at a right angle to the 1/4-inch holes. See Fig. 5-15 for a diagram of this piece.

Once these holes are drilled, saw a slot along the length of the EMT through the center of the 1/2-inch hole. File all of the holes and the edges of the slot smooth. Insert the 2-inch 1/4-20 bolts through the 1/2-inch holes so that the head of the bolt is inside the section of EMT. Thread a nut onto the bolt and tighten against the EMT. Measure the distance from the top of the head on the bolt to the center of the 3/16-inch holes in the EMT. This will be about 3/16-inch depending on the bolt. Drill two 5/32-inch holes this distance from the end of the vanes of the secondary holder and about 3/8-inch from the edge.

Insert the vane into the slot on the EMT assemblies you just made. Put a 1 1/4-inch 8-32 screw into each of the 3/16-inch holes in the EMT and through the holes in the ends of the vanes. See Fig. 5-16 for details. Tighten a nut on these screws until the slot in the EMT just closes on the vanes.

Now cut a plywood disk of the same diameter as the minor axis of the secondary mirror to serve as a backing for the secondary mirror holder. Drill a 3/8-inch hole through the center of the disk and three 1/4-inch holes equally spaced around the disk in a circle 1 inch larger than the diameter of the EMT at the center of the

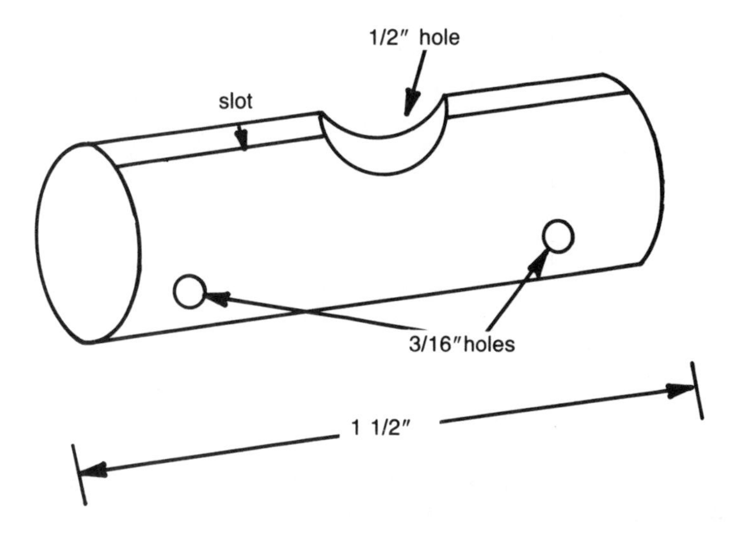

Fig. 5-15. The ends of the vanes for the secondary mirror holder.

spider. Insert the 10-24 tee nuts into these holes on the side that will face the spider assembly.

Test assemble these three parts—the mirror pad, the spider assembly, and the backing disk—as shown in the photograph in Fig. 5-13. Measure distance A in Fig. 5-17. Approximate values are shown in Fig. 5-17. Yours may vary slightly. Obtain a 1/4-20 carriage bolt 3/4-inch longer than this dimension and a washer. Place the washer into the countersunk part of the octagon and thread the carriage bolt through the octagon, spider assembly, and the plywood disk. Make sure that the tee nuts are in the side of the plywood disk that is facing toward the body of the mirror pad. Hold these pieces together by placing a washer and a wing nut on the end of the bolt. Before gluing the mirror on, put some silicon caulking around the head of the bolt to hold it in position whenever you need to adjust your mirror holder. Insert the 3 1/2-inch 10-24 screws through the disk and just tighten them against the metal plate on the octagon.

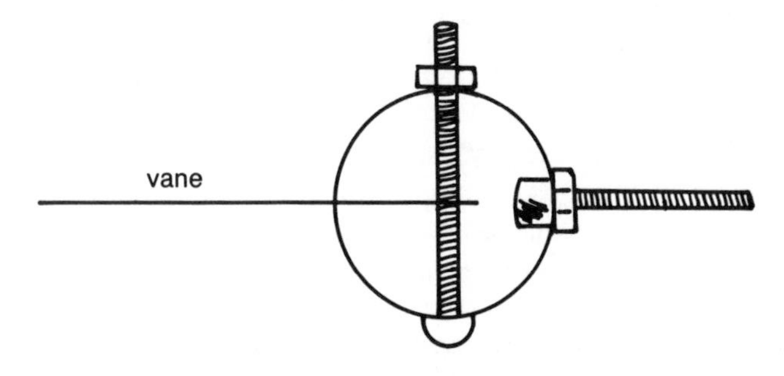

Fig. 5-16. Ends of the vanes on the spider of the secondary mirror holder.

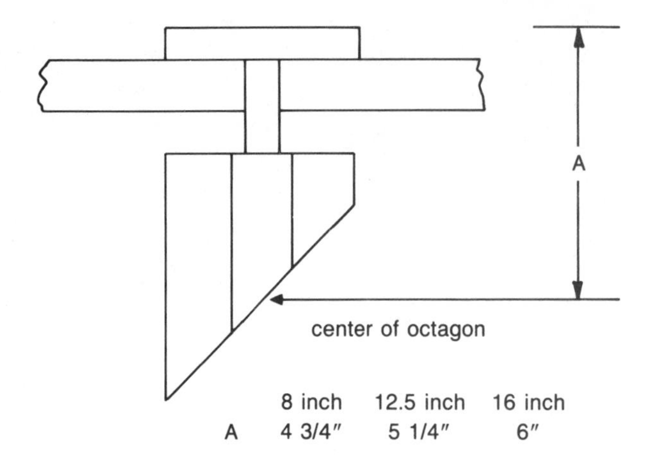

Fig. 5-17. Assembly of the secondary mirror holder.

Mount this whole assembly into the plywood octagon containing the focuser hole. Determine where to mount it by measuring the vertical distance from the center of the bolts at the ends of the vanes to the end of the head of the bolt in the octagonal block. Drill four 1/4-inch holes this distance from the center of the focuser (see Fig. 5-18) in the appropriate sides of the octagon. The plywood octagon must be positioned so that, when the secondary mirror is attached, its center will be at the center of the hole cut for the focuser. Before drilling into the octagon, cut a section of 2-×-6 just long enough to fit between opposite sides of the inside of the octagon. Use this to back up any drilling holes to prevent splintering of the plywood. Insert the bolts at the ends of the vanes into these holes. Place a washer and nut onto each of these items and tighten. See photograph in Fig. 5-19 for details of attaching the ends of the vanes to the octagon. Make sure that the center of the assembly is centered in the plywood.

Temporarily install the focuser and look through it toward the diagonal section of the secondary mirror holder. Your eye should be at least 10 inches away from the focuser and in direct line with the bore of the focuser. If you have correctly placed your secondary mirror holder in the plywood octagon, the cutout area for the bolt head will be in the center of the focuser's field of view. If it is not, make the necessary adjustments to place it there.

Remove the focuser and paint the interior of the plywood octagon and the entire secondary mirror holder and spider assembly flat black. Do not paint the face of the octagonal block to which you will be attaching the secondary mirror. Paint the exterior of the octagon whatever color you choose. It is important that this assembly be painted before installing the optics to prevent any splatters on the optics.

Mount the secondary mirror by cutting a piece of felt to fit on the diagonal end of the octagonal block. Then using aquarium ce-

Fig. 5-18. The completed secondary mirror holder.

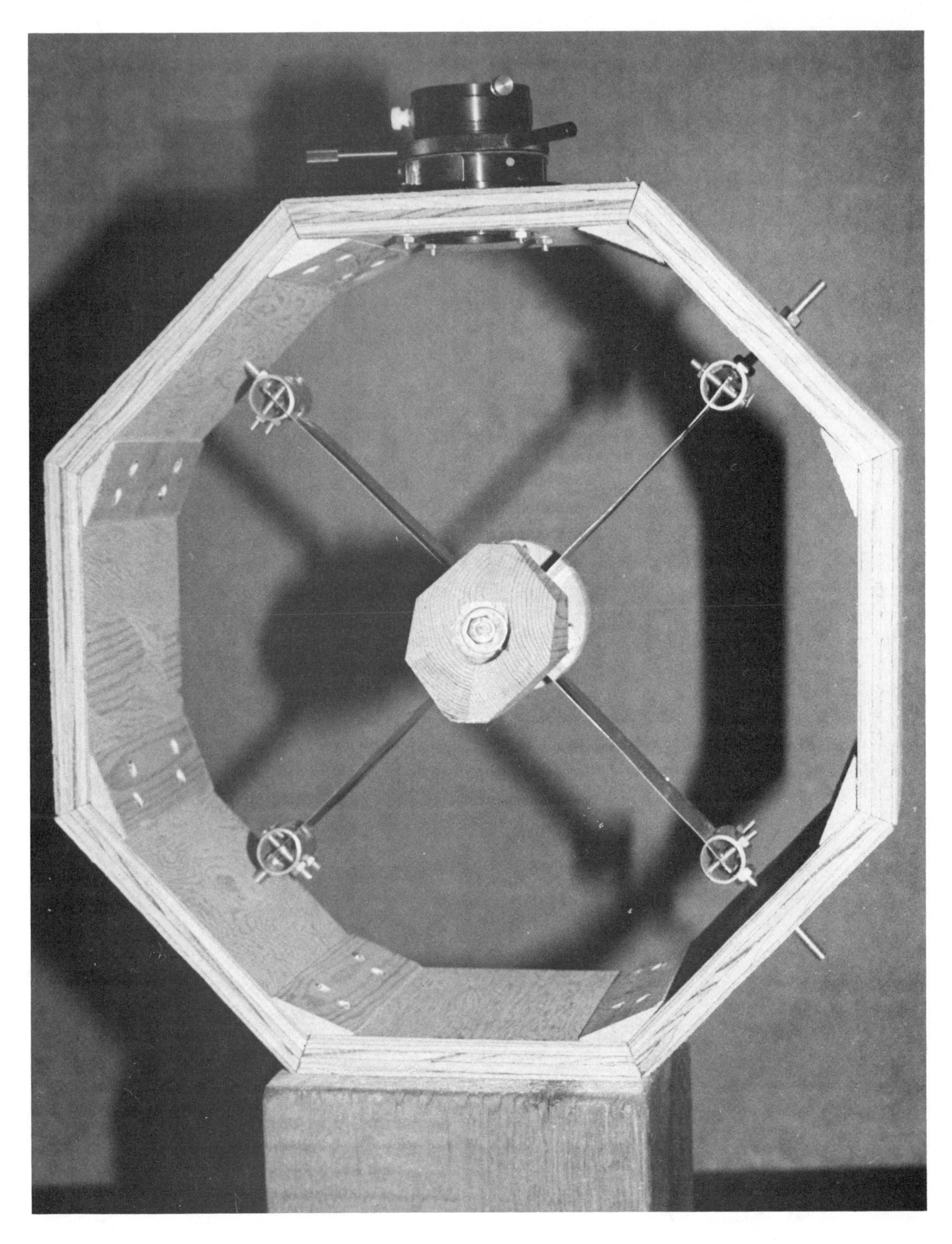

ment, glue the felt to the mirror holder and the mirror to the felt. Permanently mount the focuser using 1-inch long 10-32 screws.

ASSEMBLING THE TELESCOPE TUBE

Now that you have completed both the primary mirror-cell and the secondary mirror holder, you are ready to connect all the parts of the telescope tube. The tube for this telescope is an open tube with octagonal shaped ends to hold the mirrors and eyepiece, the trunnion box in the center to allow the tube to pivot, and lengths of EMT in between. Once all the various parts of the tube are assembled, your telescope will be finished.

First, drill 1/4-inch holes into the 6-inch wide plywood octagon that will hold the primary mirror-cell. These holes should match the locations of the 1/4-20 threaded inserts in the backup plate of the mirror-cell. Also, place these holes so that the backup plate is even with the rear of the octagon. Remember to use a board to backup the holes to prevent splintering.

Next, take the short sections of the EMT with the flattened ends and connect one end of the trunnion box to the octagon which will hold the primary mirror-cell. You should do this by first attaching two of the EMT sections to form opposite sides of the telescope tube with 1 1/2-inch long 3/8-18 bolts and nuts. Then work your way around the octagon and trunnion box, by placing additional sections of the EMT on opposite sides until you have attached all eight of them. Repeat the process with the long sections of the EMT and the 3-inch wide plywood octagon.

When all the other work has been completed, install the mirror-cell and mirror into the proper octagon and slip the secondary mirror and spider onto the front octagon.

You will find that the telescope is not completely balanced at this time. Most likely the primary mirror will be the heavier end. We found it necessary to add about eight pounds of weight distributed on the front end of the trunnion box. For this you will need to experiment to get the correct weight. You have attached the correct weight when the telescope can be pointed in any direction and has no tendency to move from that position. See Fig. 8-3 for a photograph of the finished telescope.

COLLIMATING THE OPTICS

First make a collimating eyepiece. See Fig. 5-20. This simple eyepiece consists of a 1 1/2-inch piece of 1 1/4-inch tube—a 1 1/2-inch long piece of 1 1/4-inch drain pipe works well—and a disk with a hole in it inserted into one end. Using a hole saw with a 1/8-inch drill bit, cut a disk of 3/4-inch plywood that will fit tightly into the pipe. Insert the plywood disk 5/8-inch into the pipe. Bevel the edges

Fig. 5-19. Attaching the vanes of the spider to the octagon.

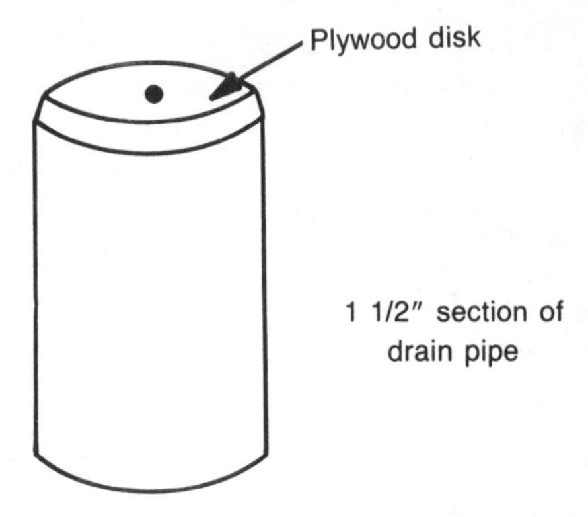

Fig. 5-20. Assembly of the collimating eyepiece.

of the plywood slightly. Paint the plywood and inside of the pipe black. An alternate method would be to obtain a section of metal dowel 1 1/4-inch in diameter and drill a 1/8-inch hole exactly through the center of the length of this metal dowel. Make sure you attach a retaining ring to the top of this dowel to prevent it from slipping through the focuser and damaging the optics.

Some amateur telescope-makers put a 1/4-inch diameter red dot in the center of the primary mirror as an aid to collimation. If you decide to do this, use red nail polish for your dot. To be sure that you get the dot in the exact center of the mirror, carefully cut a disk of cardboard exactly the size of your mirror. Punch a small hole in the center of the disk. If you draw the disk with a compass, the center is automatically marked. Place the cardboard over the reflecting surface of the mirror and line up the edges of the cardboard with the mirror. Reach through the hole in the center of the cardboard with the brush from the nail polish and place a dot of polish on the mirror. Remove the cardboard and enlarge the dot to about 1/4-inch diameter. Using a similar technique, the center of the secondary mirror can be marked with a 1/8-inch diameter dot. You need not worry about degrading the image produced with the telescope because these dots will cause less than a one percent loss of light.

Double check the position of the secondary mirror holder to be sure that it is exactly in the center of the telescope tube. If not, then make the necessary adjustments to center it.

When the secondary mirror holder is exactly in the center of the telescope tube, remove the primary mirror-cell from the telescope tube. Place a white or other light-colored object behind the telescope tube—a sheet or a piece of poster board works well. Insert the collimating eyepiece into the focuser and extend the focuser as far out as possible. The image you see should show the outline of the rear of the telescope tube centered in the image on the mirror. If this is not what you see, adjust the position of the secondary mirror until it is what you see.

To adjust the secondary mirror's position, first slightly loosen the wing nut on the mirror holder, then appropriately adjust the three 10-24 screws and tighten the wing nut again. Continue adjusting until the image of the rear end of the telescope is centered in the secondary mirror.

Return the primary mirror to its position in the telescope tube. If you placed the collimating dots on the mirrors, you simply align the mirrors by aligning these dots. Align by adjusting the three 1/4-20 screws on the primary mirror-cell. Skip the next paragraph.

If you did not paint the dots onto your mirrors, the alignment is accomplished as follows:

Point the telescope toward an evenly illuminated bright object—the blue sky works well. Using the three 1/4-20 adjustment screws at the rear of the telescope adjust the primary mirror's position until the view through the collimating eyepiece is like that in Fig. 5-21.

The final adjustments in collimating the telescope are done using the stars as the guide. Set up the telescope outside and focus on a star field, choosing a section of the sky where the stars are not too crowded. Insert a low power eyepiece and check the star images from the center out towards the edges of the mirror. If you have a good alignment, these star images will be pinpoints of light

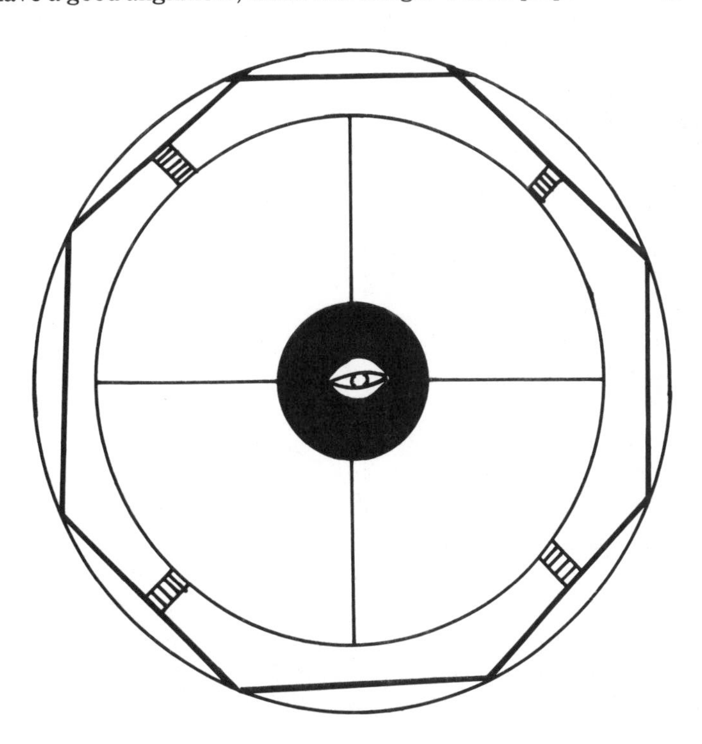

Fig. 5-21. Appearance of a collimated telescope looking through the collimating eyepiece.

across the entire field of view. If not, then make minute adjustments to the position of the primary mirror until they are.

Congratulations, your telescope is ready for use! Step back and enjoy the results of your craftsmanship and hard work. You can choose to stop here and enjoy the telescope or to continue and add the computer-controlled tracking drive.

Chapter 6

Installing the Stepper Motor Drives

NOW THAT YOUR TELESCOPE IS COMPLETE, YOU ARE READY to install the stepper motor drives for controlling the *right ascension* and *declination* movements of the telescope. Right ascension and declination are how the position of various celestial objects are noted in the sky. This corresponds to the way we determine the location of places on the earth by using longitude and latitude. To visualize this, imagine that the sky is an enormous sphere with the earth in its center and the position of all the stars is on that sphere. Then think about how we have charted out the earth in imaginary lines of longitude and latitude.

The celestial sphere has also been charted out in the same manner. Right ascension corresponds to longitude and declination corresponds to latitude. Knowing the declination and right ascension of a particular star can help you find it in the sky. Once you have set up your telescope, you can enter the right ascension and declination values of a stellar object and the computer will direct the stepper motors to move the telescope tube to that position. The computer will then direct the stepper motors to track that object as it seems to move westward across the sky.

If you have not yet mounted the fork arms and are referring to this chapter for the necessary information to do that, then move down to the section on installing the declination drives.

INSTALLING THE RIGHT ASCENSION DRIVE

This drive is the one the computer will use most frequently; it is the one which tracks objects across the sky. To install the right ascension drive, cut a hole through the brace on the U-part of the base. This hole should be in the right-hand end as you are looking from the south end of the base. The hole should be a snug fit for the stepper motor and at an angle approximately equal to:

90 minus your latitude

This is the same angle you used when installing the polar cone and should be parallel to the axis of the polar cone. For our telescope, this hole is about 2 1/4-inches in diameter and at an angle of about 45 degrees. See Fig. 6-1.

Cut three 1-inch diameter disks and two 10-inch diameter disks from A-B plywood. Drill a 3/8-inch hole in the center of two of the 1-inch disks and into the center of both of the 10-inch disks. In the center of the other 1-inch disk, drill a hole that will fit tightly onto the shaft of the stepper motor. Usually this is a 1/4-inch hole.

Put this 1-inch diameter plywood disk onto the shaft of the stepper motor and drill a 1/16-inch hole through the edge of the disk and through the shaft of the motor. Into the hole insert a 1/16-inch by 3/4-inch long roll pin to hold the disk in place on the shaft. Mount the motor into the base, positioning the stepper motor so that the face of the plywood disk is in line with the face of the polar cone. If the brace is positioned correctly, you should be able to insert 1 1/4-inch screws into the top two holes of the motor and tighten. Using a short section of 3/16-inch interior-diameter spacer tube, which just fits between the motor and the base, insert a 2 1/2-inch screw into the hole of the motor and through the spacer tube. Tighten all four screws. If this can not be done, use 3/16-inch interior-diameter spacers and the necessary lengths of screws.

Glue the other two 1-inch plywood disks to the 10-inch plywood disks such that the holes in the centers of the disks are lined up. While they are drying, drill two 3/8-inch holes through the brace on the base. One hole should be approximately 6 inches to 8 inches from the stepper motor towards the center of the brace. The other should be approximately 6 inches from the other end of the brace. Each of these holes should be drilled at the same angle which you drilled the cut for the stepper motor and close to the top of the brace.

Cut two pieces of 3/8-inch threaded rod about 8-inches long and insert them into the holes that you drilled in the brace. Then secure the rods in position by placing a washer and nut on both sides of the brace and tightening. Make four Teflon washers 5/8-inch square by 1/8-inch thick. To make the Teflon 1/8-inch thick, slice a section of the 1/4-inch Teflon in half. Drill a 3/8-inch hole in the

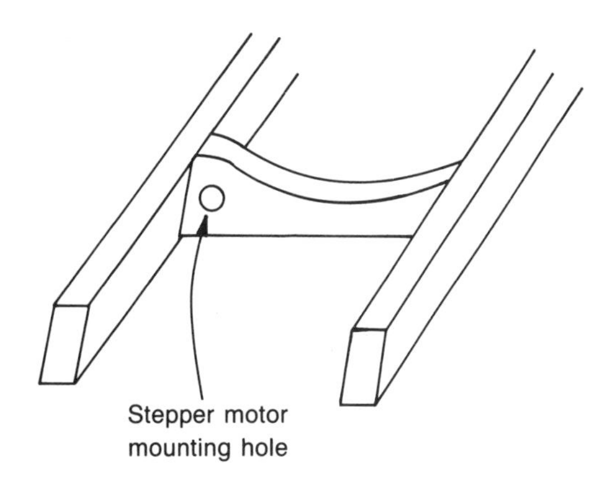

Fig. 6-1. Placement of stepper motor for right ascension drive.

center of each washer.

Place two nuts, then a Teflon washer, on the threaded rod in the center of the brace. Put one of the disk assemblies onto this rod with the 1-inch disk toward the body of the brace. Adjust the position of the disks so that the 10-inch disk is exactly in line with the disk on the stepper motor. When this is done, lock the two nuts together. Then place a Teflon washer and pair of nuts on the outside of the disks, and tighten these nuts so that the disks move freely, but with no wobble.

Repeat the above process, attaching the other pair of disks to the second threaded rod. This time the 10-inch disk should be on the side towards the brace. Line up the 1-inch disk with the disk on the stepper motor and the 10-inch disk on the center pair. When this is done, lock the disks in place.

Once the disks are in place, you will need to obtain three belts. A good type of belt to use is one made of the same material as an automobile fan belt. Many auto stores can assist you in obtaining the correct lengths. You will need a belt to reach around the stepper motor disk and the center 10-inch disk, another to fit around the center 1-inch disk and the other 10-inch disk, and finally one which will go from one end of the polar cone to the other. Figure 6-2 shows you the placement of these belts. Although the belts need to be taut, they should not place too much tension on the telescope.

INSTALLING THE DECLINATION DRIVE

Since the declination stepper motor drive is mounted onto one of the fork arms, it is easier to do the work necessary for its installation before the fork arms are mounted onto the polar cone. After you have prepared the mounting for the stepper motor, proceed with putting the fork arms onto the polar cone.

The declination stepper motor drive is to be mounted on the right-hand fork arm looking from the south end of the polar cone

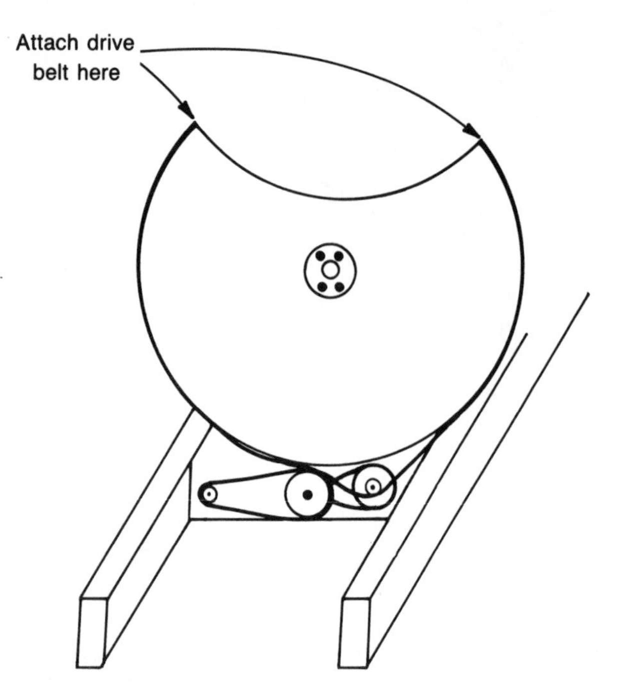

Fig. 6-2. Belts for the right ascension drive.

with the shaft of the motor projecting from the inside of the fork arm. Cut a hole through the lower right-hand corner of the appropriate fork arm. This hole should be just large enough for a snug fit for the stepper motor. For our motor, this hole is 2 1/4-inches in diameter. Cut this hole so that the edge is about two inches from the base of the fork arm and centered between the edge of the fork and the braces on the arm. Do not come closer than one inch from the side of the arm. The approximate location is shown in Fig. 6-3.

Our stepper motors have a square mounting plate that is 1/4-inch thick by 2 1/4-inch square. If your motor also has a similar mounting plate, use a chisel or a router to cut out an area around the mounting hole so that the mounting plate fits flush with the surface of the inside of the fork arm. When this is complete, return to Chapter 4 to mount the fork arms onto the polar cone.

Cut two 1-inch diameter disks and one 10-inch diameter disk from A-B plywood. Into the center of one of the 1-inch disks drill a hole sized to fit tightly onto the shaft of the stepper motor. Usually this is 1/4-inch. Drill 3/8-inch holes into the centers of the other two disks. Using contact cement, cover one side of the 10-inch diameter disk with laminated plastic and weight it for 2 to 3 hours.

While you are waiting for the glue to dry, slice two 1-inch squares of Teflon to a thickness of 1/8-inch. You will need three pieces of Teflon 1-inch square by 1/8-inch thick. Glue the 1-inch disk with the 3/8-inch center hole to the wooden side of the 10-inch disk, carefully lining up the 3/8-inch holes.

Put the other 1-inch plywood disk onto the shaft of the stepper motor and drill a 1/16-inch hole through the edge of the disk and through the shaft of the motor. Insert a 1/16-inch by 3/4-inch long roll pin to hold the disk in place on the shaft. Mount the motor on the fork arm.

To attach the 1-inch and 10-inch disk assembly onto the fork arm, drill a 3/8-inch hole at least 5 1/2 inches from the other edge and at least 5 1/2 inches from the base of the fork arm on the side of the fork arm away from the stepper motor. The approximate location is shown in Fig. 6-4. Arrange and attach the three pieces of Teflon, cut side down, symmetrically on the inside of the fork arm about three inches from this hole.

From a scrap of 1/8-inch thick Teflon, make a washer about 5/8-inch square with a 3/8-inch hole in the center. Position the glued disks onto the fork arm with the laminated plastic side against the Teflon attached to the fork arm and the Teflon washer against the 1-inch disk. Bolt the disks in place using a 3/8-inch by 4 1/2-inch carriage bolt. Make sure that the disks turn freely, but have no wobble.

Place a drive belt made from automobile fan belt material around the 1-inch disk on the stepper motor and the 10-inch disk. See Fig. 6-4 for the placement of the belt. Then place another belt of the same material around the 1-inch disk attached to the 10-inch disk and the 10-disk on the trunnion box. Carefully measure the length of belt material you need. An automobile parts store can help you get the right length.

Drill a 3/8-inch hole through the large disk of the polar cone at the base of the fork arm near the stepper motor. Feed the wires

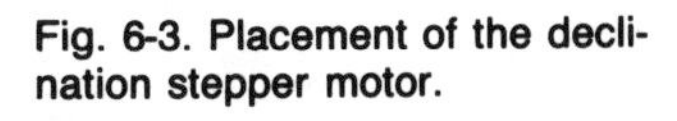

Fig. 6-3. Placement of the declination stepper motor.

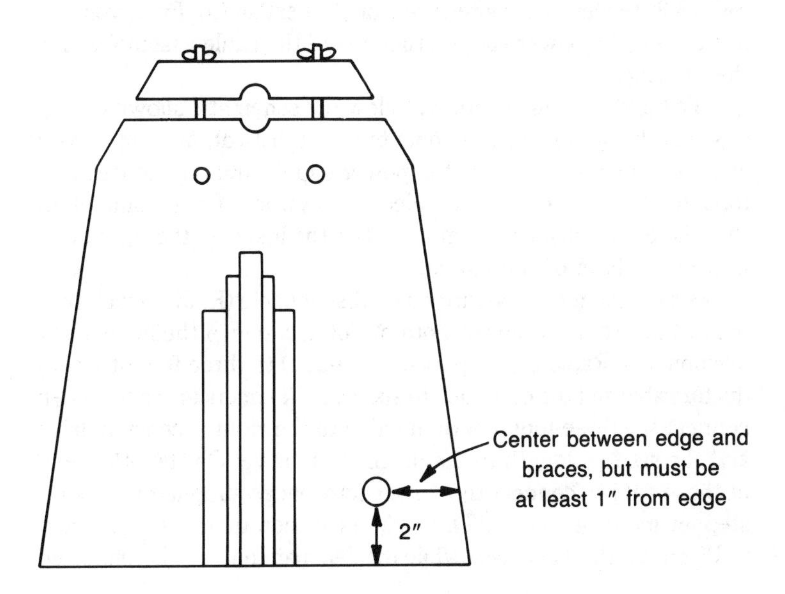

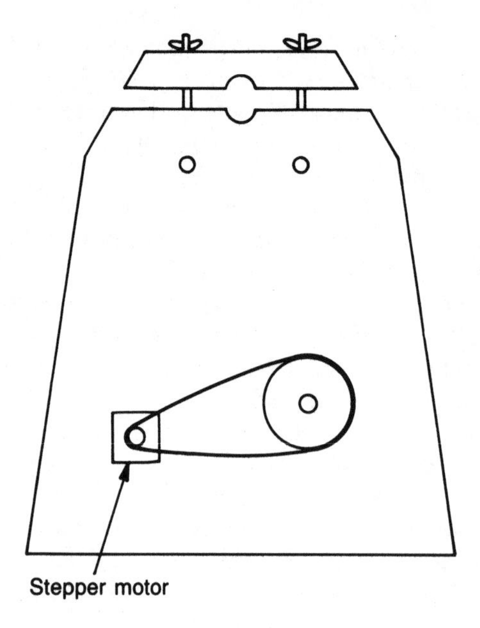

Fig. 6-4. Placement of the drive disks for the declination drive.

from the stepper motor through this hole. This stepper motor is ready to be hooked up to the interface.

HOOKING UP THE STEPPER MOTOR INTERFACE

You are now ready to hook up the stepper motor interface to the power supply and stepper motors. The purpose of the interface is to interpret the commands from the computer for use by the stepper motors. The power supply provides the power for both the interface and the stepper motors. This section refers to the use of the HS-3 stepper motor interface from CyberPak Co. First, you will need to build a power supply, then make the cable assemblies for the interface.

To build the power supply, follow the schematic shown in Fig. 6-5. The layout of the components is not critical, as long as you avoid short circuits. Mount the power supply and the stepper motor interface on a 6- × -8 inch piece of plywood. Once completed, the whole assembly can be mounted on the inside of the right hand arm of the base of the mount.

Attach the female 4-wire quick disconnects (Radio Shack part no. 274-234) to the stepper motors. Connect one of the male quick disconnects (Radio Shack part no. 274-224) to three feet of 4 conductor cable and the other one to six feet of 4 conductor cable. Then connect the three-foot length of cable to the right ascension drive and the six-foot length to the declination drive. On the other end of these cables, connect the 4 connector jacks supplied with your stepper motor interface. The right ascension motor is connected to P3 on the HS-3 interface while the declination motor is connected to P4. See Fig. 6-6 for details. The colors listed in this figure are from the stepper motor from CyberPak. If any other motor is used, then pins 1 and 2 are from one motor winding and 3 and 4 are the other. If, when you get the software running, the motors are turning the wrong way, try reversing the pairs of wires, one pair at a time, for the two windings.

Now make a cable to connect the computer and the interface. You will need a 24-pin card connector for the computer and a 26-position header for the interface. As these are not available from Radio Shack, check other electronic parts outlets in your town or a mail order company. Figure 6-7 shows the wiring needed to connect these two connectors. We used a 10-foot length of 10-wire ribbon cable to make this one. Most connectors have the letters or numbers shown in Fig. 6-7 molded onto or printed next to the wire lug.

The stepper motors are now installed. In the next chapter you will get the software up and running.

FUTURE MODIFICATIONS

You probably realize that the plywood disk drive wheels are not the most ideal for this purpose. However, in keeping with our philosophy of making this a simple telescope to build, we have side

Fig. 6-5. Power supply schematic.

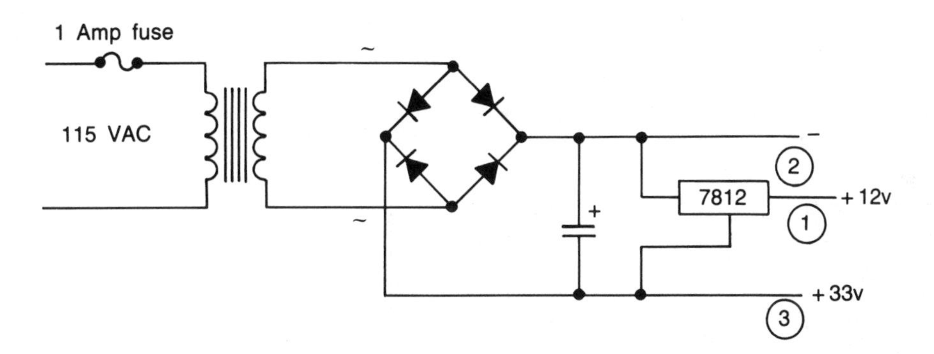

Parts list with Radio Shack Catalog numbers

Power transformer	25.2v 1.2A		273-1353
Fuse holder			270-739
Fuse 1A			270-1284
Bridge rectifier	4A	100PIV	276-1171
Capacitor	4700MF	35V	272-1022
Voltage regulator	7812		276-1771

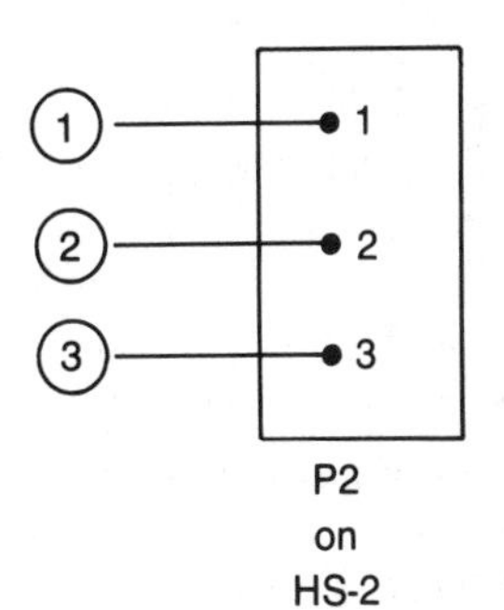

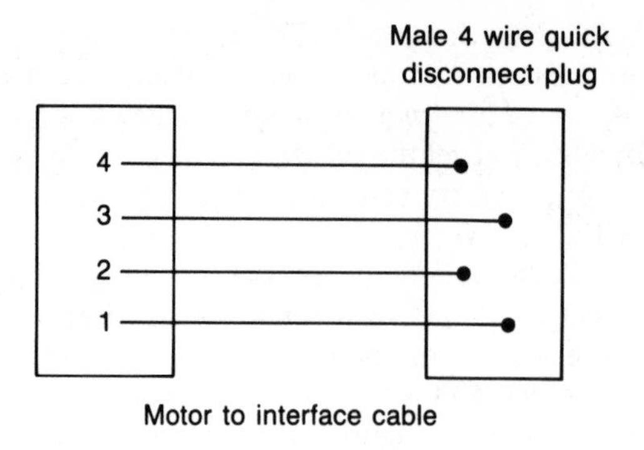

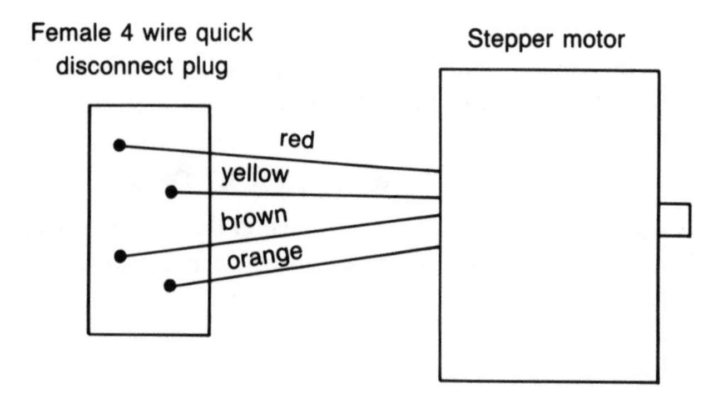

Fig. 6-6. Cables from the stepper motor to the interface.

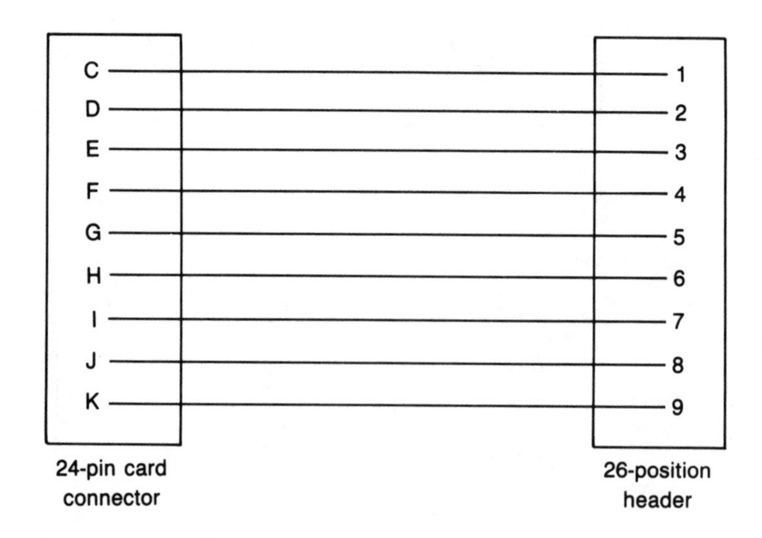

Fig. 6-7. Wiring cable between the computer and interface.

stepped the idea of making a gear train or chain drive system for this drive system.

Once you have become comfortable with this telescope, you will probably want to change step intervals on the stepper motor and move to a gear or chain drive system. For more ideas on this, we recommend that you read the book *Microcomputer Control of Telescopes* by Trueblood and Genet (see Appendix D). However, be prepared to experiment to get the right drive system for your telescope.

Chapter 7

Installing the Control Software

YOU ARE NOW READY TO TYPE IN THE SOFTWARE PROGRAMS that will control your telescope. If you have obtained the diskette with the software on it, skip the rest of this section. If not, follow these instructions carefully.

Appendix A contains three program listings, Listings 1 and 2 are the BASIC programs used to control the telescope. These are the programs which you are to type in. Listing 3 is the assembly language source code for the machine language subroutines in Listing 2. This listing is included to allow you to add features to the program.

When typing in Listings 1 and 2, be careful that you enter the programs exactly as they are listed. Any typographical errors will probably prevent the programs from executing correctly. After you type in each listing, save it onto a diskette before running it. When saving Listing 1, call it by the name CONTROL. This program loads a set of machine language subroutines called STEP.EX. Save Listing 2, calling it whatever you like. The program in Listing 2 will create the STEP.EX file on the disk. If you made an error in entering the numbers in the DATA statements, the program will inform you that an error exists. If you get an error message, then compare your program with the one printed in this book to find the error.

Once you have typed in and saved both Listing 1 and Listing

2, RUN the program in Listing 2. The software is now ready for use.

OPERATING THE SOFTWARE

Load the control program by typing:

LOAD "CONTROL" ,8

When it has finished loading, type RUN and press the RETURN key. A title page will be displayed as the machine language subroutines are being loaded and the constants and some other items are being initialized.

The screen will then clear and the question:

Enter the current date as
HH,DD,YYYY?

with a flashing cursor beside the question mark will be displayed. Enter today's date. The computer will not allow dates before 1985. You can enter the years 1985 to 1999 using the last two digits (for example, enter 1987 as 87), but any other dates must be entered as 4 digits.

The next several questions will start the time in the clocks. The computer keeps track of and displays three different times—Universal Time, Standard Time, and Local Time. These times are important because the computer uses them in various calculations. To help the computer be accurate in these calculations, enter the starting time as close to the actual time as possible. If you are in an area using Daylight Savings time, subtract an hour. When answering the first question, which is the starting time for the clocks, add 30 seconds to the starting time to allow yourself enough time to answer the other questions.

The first question the computer asks to start the clocks is:

Please enter the clock start time
As HHMMSS?

For example, the answer 013230 is equivalent to the time 1:32 and 30 seconds.

The computer will then ask:

Is this time AM or PM?

You should respond with either A or P. The computer considers midnight as AM and noon as PM.

The next two questions ask for constant values. The first is:

What is your longitude?

You can obtain your approximate longitude from a state road map, or for a more accurate value, from topographic maps available from the U.S. Geological Survey or your local library. If possible, get the value to the nearest 0.1 degree.

This is followed by the question:

What is your time zone (0-23)?

For the United States the time zones are Eastern—5, Central—6, Mountain—7, and Pacific—8. Some countries use fractional time zones. The computer will not accept fractional time zones. Your entry must be a whole number.

You can answer these last two questions each time you load the program, or because the value of these questions will not change as long as you are setting up your telescope in the same location, you can enter new lines in the CONTROL program like this;

```
1225 LN=your longitude: GOTO 1255
1255 TZ=your time zone: GOTO 1320
```

In line 1225, substitute your longitude for the words "your longitude" and in 1255, substitute your time zone for the words "your time zone". Then save the program using the following procedure:

```
OPEN 1,8,15,"S0:CONTROL": CLOSE 1
```

When the flashing cursor reappears then type:

```
SAVE "CONTROL" ,8
```

Now when the cursor reappears, the changes will be on your disk, and you will no longer have to answer these questions each time you set up your telescope-computer system for use.

When you have finished answering these questions, there will be a slight pause and the computer will display:

Press space to start clocks.

Be sure that the clock time is the same as the time you entered above. When it is, press the space bar. At that point, the clocks will begin running. Later they will be displayed on the screen.

STARTING THE TELESCOPE TRACKING

With the preliminaries finished, you can now get down to the business of actually controlling the telescope. The computer is waiting for you to set a reference point with the telescope. Set this reference point by pointing the telescope to a bright star. When the bright star is centered in your eyepiece, change to a higher power eyepiece and center it again.

Once the star is centered in a high powered eyepiece, press the space bar on the computer keyboard. The telescope will immediately begin tracking the star, and you will be asked to enter the coordinates of the star. First the computer will ask:

ENTER THE DECLINATION (AS +/-DD.D)?

This simply means that you enter the declination in degrees using the form of two digits and one decimal digit preceeded by either a plus (+) or a minus (-) sign. When the + sign is appropriate, enter it; do not assume it. The valid range for the declination is +90.0 to -90.0, but of course only a portion of this range will be visible to you at a given time on a given date. You will need to know what the declination is for the star you are viewing, as the computer does not calculate the horizon. Entering the wrong numbers could cause the eyepiece to crash into the body of the telescope.

Once the declination is entered and checked, the computer will ask:

NOW ENTER THE RIGHT ASCENSION OF THE
OBJECT (AS HH,MM)?

While the valid range for HH is 0 to 23 and for MM is 0 to 59, only a portion of this will be visible to your telescope at a given time on a given date. The computer will calculate what can be seen. The right ascension of the brightest stars visible from approximately 40 degrees north latitude are given in Appendix B.

The computer will allow you to track objects within four hours of right ascension of local zenith. This is two-thirds of the distance to the horizon. Seldom will you want to track closer to the horizon than this because of atmospheric haze. If you do wish to track closer to the horizon, change the numeral 4 in program lines 3550, 4110, and 4120 to either a 5 or a 6. The number must be a whole number. The maximum value that can be entered here is 6. A 6 will allow you to move the telescope from horizon to horizon. Also change the numeral 8 in line 2020 to twice the number entered in the three lines mentioned above.

THE CONTROL MENU

Now the control menu, along with some other information, is displayed on the screen. The screen looks is shown in Fig. 7-1.

At the top of the screen are the three running clocks. The clock on the left is the Universal Time clock. Universal Time, sometimes called Greenwich Mean Time, has been adopted as the standard time for astronomical purposes and is the time at the prime meridian at the Royal Observatory in Greenwich, England.

The middle clock is the Standard time for your time zone. If you are using Daylight Savings time, it will be an hour behind local clock time. This clock is a 12-hour clock with noon considered PM and midnight considered AM.

The clock on the right is your local time. Most observers are not located at the meridian or the center of the Standard time zone. Thus local time differs somewhat from the Standard time. For example, the center of the Eastern time zone in America is 75 degrees west longitude. This runs a few miles to the east of Philadelphia. But if you live in Detroit, your longitude is 83 degrees west. That difference of eight degrees means that the moon will rise about a half hour later, Standard time, in Detroit than in Philadelphia. Moonrise will be the same time on the local time clock for both cities.

Just below that is the date. It will change at midnight local Standard time. The computer will correctly change the date even for a leap year and New Year's Eve.

Under the date are the current coordinates for the telescope

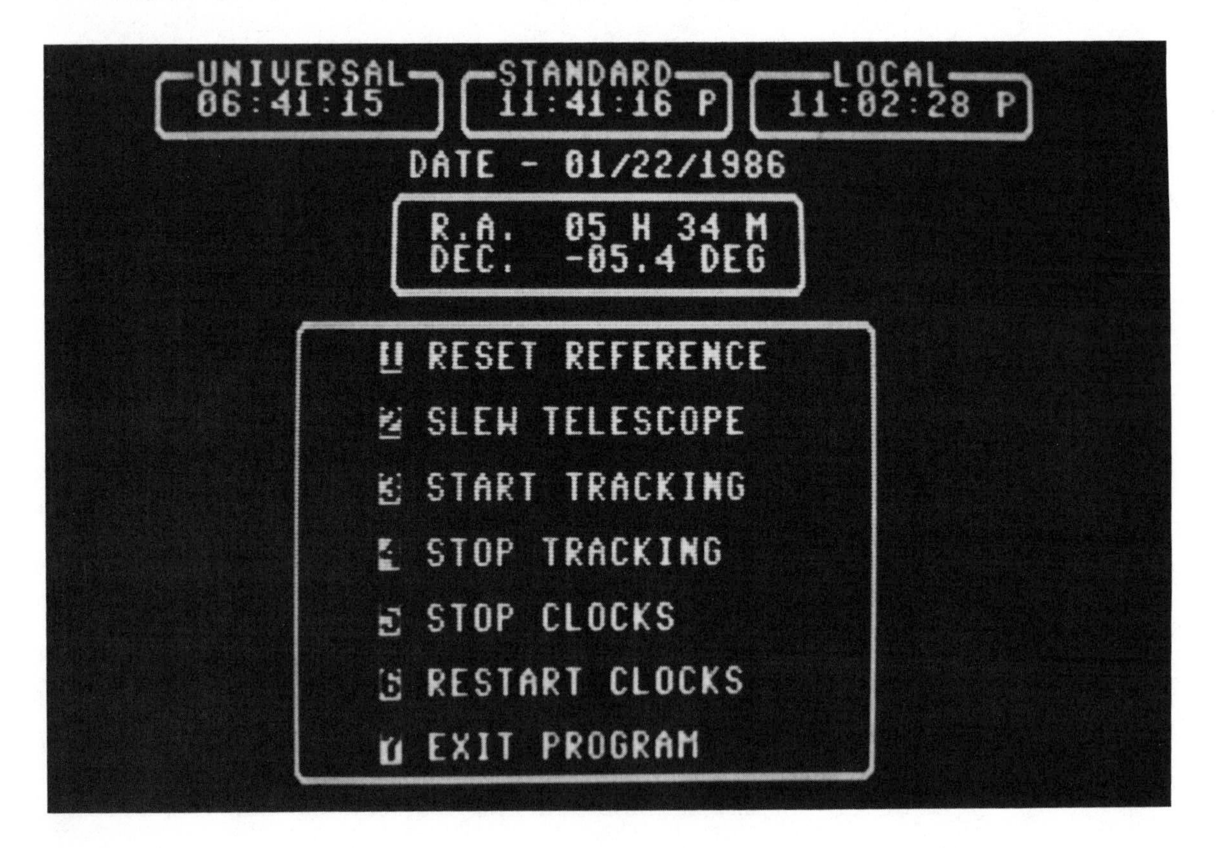

Fig. 7-1. Screen displaying the clocks and the control menu.

system. This tells you the current declination and right ascension at which the telescope is pointing. You can change these coordinates three ways. Two are via the first menu options described below; the third is by using a joystick connected to the computer. The joystick, connected to control port 2 on the computer, will allow fine movement of the telescope. When you have moved a certain number of steps in either axis, the display will update the coordinates.

Finally, the control menu is listed. When you select a menu item, an asterick (*) will appear beside the item you selected. It may only flash briefly as some menu items do not take long to execute and do not otherwise affect the screen. Here is a list of each of the menu items with a description of each:

• ***Reset Reference.*** This option allows you to point the telescope toward a new bright star and reset the reference point. To do this, you follow the same steps that you did earlier.

Once the new star is centered in a high powered eyepiece, press the space bar on the computer keyboard. The telescope will immediately begin tracking the star, and you will be asked to enter the coordinates of the star. First the computer will ask:

ENTER THE DECLINATION AS +/-DD.D ?

This simply means that you enter the declination in degrees using the form of two digits and one decimal digit preceeded by either a plus (+) sign or a minus (−) sign. Do not assume the + sign. The valid range for the declination is +90.0 to −90.0, but you must check to make sure the numbers are correct for your location, as the computer does not calculate the horizon. Incorrectly entered numbers could cause the eyepiece to crash into the body of the telescope.

Once entered and checked, the computer will ask:

NOW ENTER THE RIGHT ASCENSION OF THE
OBJECT AS HH,MM ?

The valid range for HH is 0 to 23 and for MM is 0 to 59, but as only a portion of that is visible at a given time on a given day, enter the appropriate numbers.

• ***Slew Telescope.*** You can have the telescope move to a new set of coordinates using this option. First the computer asks:

ENTER THE DECLINATION AS +/-DD.D ?

Here you enter the declination as above and then the right ascension:

> NOW ENTER THE RIGHT ASCENSION OF THE OBJECT AS HH,MM ?

Once you have entered the coordinates to which you wish to move, the computer will check them for validity and then, if valid, move the telescope to them.

- ***Start Tracking.*** If you have chosen to stop the telescope tracking using the next option, you can use this option to restart the tracking.
- ***Stop Tracking.*** This option will stop the telescope tracking system. Remember, that if stopped very long, the coordinates shown on the screen will be wrong, and you will need to reset the reference.
- ***Stop Clocks.*** You may use this option to stop the time from being displayed on the screen. The clocks, however, will continue to keep time.
- ***Restart Clocks.*** This will allow the time to be displayed again on the clocks on the screen, after they were stopped by the previous option.
- ***Exit the Program.*** Use this option to exit the program. There are a number of changes made to the computer's memory which need to be restored to original condition. If you exit the program by pressing the RUN/STOP key alone or with the RESTORE key, the computer will likely lock up when running another program.

If you are a knowledgable programmer, you might wish to add features, redo those features already in the programs or otherwise improve the program. For example, the SLEW menu entry would be aesthetically more pleasing if both the axes moved together so that the telescope would seem to move in a straight line to the new location. We chose not to do this so the program would be more understandable to the less experienced programmer.

If you make major modifications or improvements to this operating system, we would like to see the changes. Send them to us in care of the publisher.

Chapter 8

Using Your Computer Controlled Telescope

CONGRATULATIONS, YOU HAVE NOW COMPLETED YOUR telescope! This chapter is designed to be your user's guide. It will give you suggestions as to the setup, use, and maintenance of your new telescope.

First, find a location to set up your telescope. It is best if this is a location that is always available to you. Once you have found such a site, determine the direction of true north. Keep in mind that a compass points to the magnetic north pole in northern Canada, not to the actual north pole. Unless you live in Western Europe or in Eastern Asia, there will be a difference between true north and the direction the compass points. This difference is given on the bottom of the topographic map you used in Chapter 2 to find your altitude, latitude, and longitude.

If possible, make permanent markings at your observation site to indicate the placement of the base of your telescope. Remember, the arms of the U-shaped base open toward the north, and the bottom of the U, with the bearing for the polar cone, faces south. When setting up your markings, use your imagination. They can be anything from a concrete slab with an outline of the telescope base painted on it to bricks buried for each corner of the base. Even better would be a permanent observatory building. Choose something appropriate for your location.

Fig. 8-1. Make sure the base is level.

SETTING UP YOUR TELESCOPE

After an observation site is found and prepared, you are ready to set up your telescope. First place the base in position, orienting it properly towards the north. Make sure that the base is level. (See Fig. 8-1) Next install the polar cone. Tighten the south bearing only enough to snugly hold the axle, but not so tightly as to prevent smooth rotation of the polar cone. (See Fig. 8-2) Experiment with this to learn the best amount of tightness. Loop the belt attached to the disk around the drive disk on the base.

Place the telescope tube onto the fork arms, making sure that your drive disk faces the arm with the stepper motor. Attaching the telescope tube to the fork arms might require some practice, as you will have to hold it in place while putting the top half of the declination bearings onto the fork arms and tightening the wing nuts. Again, experiment with how much to tighten the bearings. (See Fig. 8-3) You want them tight enough to hold the tube snugly, without restricting its free rotation. Loop the belt attached to the telescope tube around the drive disk attached to the fork arm.

Fig. 8-2. Tighten the south bearing.

Next connect the cable from the stepper motor on the fork arm to the interface, and connect the cable from the interface to the computer. Turn on the computer, video monitor, disk drive, and interface. A suggestion here—attach all of these to an outlet bar, then you can power up everything with a single switch.

When the telescope and equipment are all set up, load the CONTROL program and answer the date and time questions. Then set your reference point with the telescope. With a low-power eyepiece in the telescope, find one of the stars from Appendix B and center it. Then switch to a high-powered eyepiece and center the star again. Press the space bar on the computer keyboard, and the computer will begin tracking the star. Enter the declination and right ascension of the star.

You are now ready for a session of observing. Whenever you want the telescope to move to a new object in the sky, select option 2 from the menu on the screen. Enter the coordinates for this object and the computer will move the telescope to the new object.

If the drive disks are perfectly circular, when the telescope stops moving, the object will be in the field of vision. If not, you may need to do some hunting. For such hunting, you can look through the eyepiece and adjust the position of the telescope with a joystick.

Periodically, you may find that the small positioning errors accumulate to a larger error. An object you wish to find is not visible in the eyepiece, nor is it nearby in the sky. In a case like this, choose option 1 from the menu on the screen and set a new reference point.

COLLIMATING THE OPTICS

Periodically, you will need to fine tune the optical alignment to obtain the best performance from your telescope. The procedure to do this is outlined in the last section of Chapter 5. We find that this tune-up is required every time the telescope is transported some distance in the car. Otherwise, you should do a quick check of the alignment after every five to ten observing sessions.

CLEANING THE MIRROR

It is important to keep your mirror covered with a dustproof cover whenever you are not using the telescope. The surface of the mirror is delicate and easily scratched, so you do not want to have to clean it very often. However, a little dust is better than a little scratch.

Should you need to clean the mirror, the following method works well. Remove the mirror-cell from the telescope tube and place it on a stable table top. Carefully detach the mirror from the cell. At no time should you touch the optical surface of the mirror.

Fig. 8-3. Experiment with tightening the bearings.

Then carefully place the mirror on a soft towel in your kitchen sink, submerging it into lukewarm water containing a small amount of dish washing detergent. Ivory Liquid works best. If you have hard water or particles in your water, use bottled, distilled water.

For the following steps be sure that you have obtained high-quality cotton balls. Cheaper or generic brands often have small abrasive particles in them. Using a clean cotton ball, gently wipe the submerged surface of the mirror with even back and forth strokes. Do not use circular strokes and do not apply any pressure. Start at one side of the mirror and work to the other side. With a fresh cotton ball, again wipe the surface of the mirror using strokes at a right angle to those used the first time. Drain the water from the sink and flood the surface of the mirror with fresh water. Do not allow the surface to dry or water to stand in beads on the mirror.

After the soapy water has been rinsed from the mirror, rinse thoroughly with distilled water. Then saturate a fresh cotton ball with rubbing alcohol and wipe the mirror with even, gentle strokes, beginning at one side and moving to the other. Using another cotton ball saturated with alcohol, wipe the surface of the mirror at right angles to the strokes used the first time. Follow this up by wiping with dry cotton balls. Finally, place the mirror face down on a soft, lint free towel.

After standing for about two to three hours on the towel, the mirror is ready for reinstallation into the telescope. Follow the procedures outlined in Chapter 5 for the installation and collimation of the optics.

Appendix A

Listings of BASIC and Assembly Language Programs

FOLLOWING ARE THE LISTINGS OF THE PROGRAMS REQUIRED for the operation of the Commodore 64 computer for controlling your telescope. Listing A-1 is the BASIC control program. Next is a list of the variables used in the program along with a brief description of the function of each variable. Listing A-2 is a program that will generate a machine-language program file called STEP.EX on your diskette, and is loaded by the BASIC program in Listing A-1.

The final listing, Listing A-3, is the assembly language source code for the machine-language portion of the telescope control system. We used the MAE assembler from Eastern House Software. While this assembler has some nonstandard assembler features, it has worked well for us for a long time. If you have the diskette containing these programs, this source code is called STEP.CONT. Also included on the diskette is a file called STEP.SEQ. STEP.SEQ is the same source code translated to PETASCII format sequential file and should allow you to translate the source for use with a variety of other assemblers.

Listing A-1. BASIC Control Program.

```
100 rem  telescope control program
110 rem  written by richard f. daley
120 rem  for the commodore-64
130 rem
```

```
140 rem  copyright 1985
150 rem  permission to copy
160 rem  but not to sell or distribute
170 rem
180 if a=1 then 370
190 rem
200 rem *** display title page ***
210 rem
220 print "[illegible]";chr$(8);chr$(14)
230 poke 53280,7: poke 53281,9
240 print "[illegible]";tab(8);"Telescope Operating System"
250 print "[illegible]";tab(19);"by"
260 print "[illegible]";tab(9)"Richard F. Daley, PhD"
270 print tab(18);"and"
280 print tab(13);"Sally J. Daley"
290 print "[illegible]";tab(13);"Copyright 1985"
300 print "[illegible]";tab(14);"Published by"
310 print tab(13);"Tab Books, Inc.[illegible]"
320 print "Now loading machine language . . .[illegible]"
330 rem
340 rem *** load machine language subroutines ***
350 rem
360 if a=0 then a=1: load "0:step.ex",8,1
370 clr: for i=0 to 1500: next i
380 print "                                   [illegible]"
390 print "Now initializing memory . . ."
400 clr
410 rem
420 rem *** open logical file from keyboard ***
430 rem
440 open 1,0,1
450 rem
460 rem
470 rem *** constants for access to machine language ***
480 rem
490 tt=49152: rem local time storage buffer
500 lt=tt+3: rem local time storage buffer
510 ra=lt+3: rem steps to slew in r.a.
520 rn=ra+3: rem direction for slew
530 r1=rn+1: rem correction from joystick
540 dc=r1+2: rem steps to slew in dec
550 dn=dc+3: rem direction for slew
560 d1=dn+1: rem correction from joystick
570 sp=d1+2: rem set up stepper motor interrupts
580 cs=sp+3: rem disable stepper motor interrupts
590 sl=cs+3: rem slew stepper motors to new scope location
600 tm=sl+3: rem set up time display interrupts
610 rs=tm+3: rem restart time display interrupts
620 cl=rs+3: rem disable time display interrupts
630 ue=cl+3: rem enable freewheel mode on steppers
640 en=ue+3: rem energize steppers
650 rem
660 rem *** other constants ***
670 rem
680 sd=72: rem steps per minute in r.a.
690 ds=15: rem steps per degree in dec.
700 gm=6.6404: rem gmst at 0h 1/0/1985
710 dt=0.06571: rem daily change in gmst
720 ep=725006: rem days from 0/0/0 to 1/0/1985
730 cd$="[illegible]"
740 rem
750 rem *** set up calendar data ***
760 rem
770 dim d(12)
780 for i=1 to 12: read d(i): next i
790 data 32, 29, 32, 31, 32, 31, 32, 32, 31, 32, 31, 32
```

```
800 sys sp: sys cs: sys ue
810 for i=0 to 2500: next i
820 rem
830 rem *** get date and clock data ***
840 rem
850 poke 53280,0: poke 53281,0
860 print "■■■■"
870 print "Enter the current date as": print "MM,DD,YYYY? ";
880 input#1,mo$,dy$,yr$: print: yr=val(yr$): mo=val(mo$)
890 if yr>=1985 then 940
900 if yr>100 then 920
910 if yr>84 then 930
920 print "■Date should be after Dec. 31, 1984": goto 870
930 yr=yr+1900: yr$=str$(yr): yr$=right$(yr$,len(yr$)-1)
940 if mo<1 or mo>12 then print "■Month out of range!": goto 870
950 dy=val(dy$)
960 lp=-(int(yr/4)=(yr/4))+(int(yr/100)=(yr/100))-(int(yr/400)=(yr/400))
970 if mo<>2 then 1000
980 if dy>28+lp then print "■Day out of range!": goto 870
990 goto 1010
1000 if dy<1 or dy>=d(mo) then print "■Day out of range!": goto 870
1010 print "■■Now enter the time to start the clock."
1020 print "If you are using daylight savings time,"
1030 print "then subtract one hour. Add 30 seconds"
1040 print "for processing."
1050 print "■Please enter the clock start time"
1060 print "as HHMMSS? ";: gosub 3120: tm$=t$
1070 rem
1080 rem *** make sure time entry is valid ***
1090 rem
1100 if len(tm$)<>6 then print "■Need 6 numbers for time.": goto 1050
1110 for i=1 to 6: t=asc(mid$(tm$,i,1))
1120 if t<48 or t>57 then print "■Non number in time.": goto 1050
1130 next i
1140 if val(left$(tm$,2))>12 then print "■Hours > 12.": goto 1050
1150 if val(left$(tm$,2))<1 then print "■Hours < 1.": goto 1050
1160 if val(mid$(tm$,3,2))>59 then print "■Minutes > 59.": goto 1050
1170 if val(right$(tm$,2))>59 then print "■Seconds > 59.": goto 1050
1180 print "■Is this time ■A■M or ■P■M? ";
1190 gosub 3160
1200 b=0: c=0: if a$="p" then b=128: c=128: goto 1220
1210 if a$<>"a" then 1190
1220 print "■";a$;"■"
1230 print "■What is your longitude? ";
1240 gosub 3120: ln=val(t$)
1250 if ln<0 or ln>=360 then print "■Out of range!": goto 1230
1260 print "■What is your time zone (0-23)? ";
1270 gosub 3120: tz=val(t$)
1280 if tz<0 or tz>23 then print "■Out of range!": goto 1260
1290 rem
1300 rem *** calculate universal time ***
1310 rem
1320 t$=tm$: t1=0
1330 if a$="p" then if left$(t$,2)="12" then t1=1
1340 if a$="a" then if left$(t$,2)="12" then t$="00"+right$(t$,4)
1350 ut=val(left$(tm$,2))+tz-12*(a$="p")-12*t1
1360 if ut>23 then ut=ut-24
1370 ut$=str$(ut): ut$=right$(ut$,len(ut$)-1)
1380 if ut<10 then ut$="0"+ut$
1390 ut$=ut$+right$(tm$,4): tm$=t$
1400 rem
1410 rem *** calculate local time correction ***
1420 rem
1430 t=4*(int((tz*15-ln)*100)/100)
1440 cr=1: if t<0 then cr=-1
1450 t=abs(t): mm$=str$(int(t)): mm$=right$(mm$,len(mm$)-1)
```

```
1460 t=abs(int(60*(t-val(mm$))+.005))
1470 ss$=str$(int(t)): ss$=right$(ss$,len(ss$)-1)
1480 rem
1490 rem *** calculate days from 0 jan 1985 to date ***
1500 rem
1510 m=val(mo$): d=val(dy$): y=val(yr$)
1520 t1=0: if m=1 then 1540
1530 for i=1 to m-1: t1=t1+d(i)-1: next i
1540 t1=t1+y*365+int(y/4)+d+1-int(y/100)+int(y/400)+(m<3)
1550 nd=int(t1-ep+.5)
1560 rem
1570 rem *** set up time in clock buffers ***
1580 rem
1590 hh=val(left$(tm$,2))
1600 t$=mid$(tm$,3,2): mm=val(t$)+(val(mm$)*cr)
1610 t$=right$(tm$,2): ss=val(t$)+(val(ss$)*cr)
1620 if ss<0 then ss=ss+60: mm=mm-1
1630 if mm<0 then mm=mm+60: hh=hh-1
1640 if hh<1 then hh=hh+12: c=0
1650 if ss>59 then ss=ss-60: mm=mm+1
1660 if mm>59 then mm=mm-60: hh=hh+1
1670 if hh>12 then hh=hh-12: c=128: if b then c=0
1680 t$=str$(hh): lt$=right$("00"+right$(t$,len(t$)-1),2)
1690 t$=str$(mm): lt$=lt$+right$("00"+right$(t$,len(t$)-1),2)
1700 t$=str$(ss): lt$=lt$+right$("00"+right$(t$,len(t$)-1),2)
1710 t=tt: t$=tm$: k=b: gosub 3220
1720 t=lt: t$=lt$: k=c: gosub 3220
1730 rem
1740 rem *** start clocks ***
1750 rem
1760 print "■Press space to start clocks."
1770 gosub 3160: if a$<>" " then 1770
1780 sys tm: ti$=ut$: sys cl: ft=0
1790 gosub 3310: poke rn,0: poke r1,0: poke r1+1,0
1800 poke dn,0: poke d1,0: poke d1+1,0
1810 rem
1820 rem *** set up screen ***
1830 rem
1840 gosub 3710: sys rs: ft=1
1850 t$=ti$: print "■■";tab(2);left$(t$,2)"■|";mid$(t$,3,2);"■|";right$(t$,2)
1860 gosub 3830
1870 get a$: a=val(a$): if a<1 or a>7 then 2510
1880 print left$(cd$,2*a+8);tab(8);"■*■"
1890 on a goto 1930, 1980, 2250, 2300, 2350, 2420, 2470
1900 rem
1910 rem *** reset reference ***
1920 rem
1930 gosub 3300: gosub 3710: sys rs: ft=1: gosub 3830
1940 goto 2510
1950 rem
1960 rem *** slew telescope ***
1970 rem
1980 if ft=1 then sys cl
1990 print "■■■input desired destination coordinates."
2000 tc$=dc$: th$=rh$: tm$=rm$
2010 gosub 3370: n=val(rh$)
2020 if abs(val(th$)-n)>8 then print "■movement > 8 hrs in r.a.": goto 2010
2030 gosub 4030
2040 if n>e then if fg<>1 then print "■movement too far east!": goto 2010
2050 if (n-24)>e then if fg=1 then print "■movement too far east!": goto 2010
2060 if n<w then if fg<99 then print "■movement too far west!": goto 2010
2070 if (n+24)<w then if fg>98 then print "■movement too far west!": goto 2010
2080 mr=int(sd*((60*val(rh$)+val(rm$))-(60*val(th$)+val(tm$)))+.5)
2090 t1=1: if mr<0 then t1=255
2100 mr=abs(mr): poke rn,t1
2110 md=int(ds*(val(dc$)-val(tc$))+.5)
```

```
2120 t1=1: if md<0 then t1=255
2130 md=abs(md): poke dn,t1
2140 t1=int(mr/65536): t2=int((mr-(t1*65536))/256): t3=mr-t1*65536-t2*256
2150 poke ra,t3: poke ra+1,t2: poke ra+2,t1
2160 t1=int(md/65536): t2=int((md-(t1*65536))/256): t3=md-t1*65536-t2*256
2170 poke dc,t3: poke dc+1,t2: poke dc+2,t1
2180 sys sl
2190 gosub 3710: gosub 3830
2200 if ft=1 then sys rs
2210 goto 2510
2220 rem
2230 rem *** start stepper ***
2240 rem
2250 if fs=0 then sys sp: fs=1
2260 goto 2510
2270 rem
2280 rem *** stop stepper ***
2290 rem
2300 if fs=1 then sys cs: fs=0
2310 goto 2510
2320 rem
2330 rem *** stop clock display ***
2340 rem
2350 if ft=1 then sys cl: ft=0: goto 2370
2360 goto 2510
2370 gosub 3710
2380 gosub 3830: goto 2510
2390 rem
2400 rem *** start clock display ***
2410 rem
2420 if ft=0 then sys rs: ft=1
2430 goto 2510
2440 rem
2450 rem *** exit the program ***
2460 rem
2470 gosub 3650: print chr$(9): end
2480 rem
2490 rem *** update date at 12 p.m. local standard time ***
2500 rem
2510 t$=ti$: if val(left$(t$,2))<>tz then 2660
2520 if right$(t$,4)<>"0000" then fl=0: goto 2660
2530 if fl then 2660
2540 dy=val(dy$)+1: nd=nd+1: mo=val(mo$)
2550 if dy<d(mo) then 2610
2560 if mo<>2 then 2580
2570 if lp=1 then if dy=29 then 2610
2580 dy=1: mo=mo+1: if mo<13 then 2600
2590 mo=1: yr=val(yr$)+1: yr$=str$(yr): yr$=right$(yr$,len(yr$)-1)
2600 mo$=str$(mo): mo$=right$(mo$,len(mo$)-1): if len(mo$)=1 then mo$="0"+mo$
2610 dy$=str$(dy): dy$=right$(dy$,len(dy$)-1): if len(dy$)=1 then dy$="0"+dy$
2620 print "■■■■";tab(11);"date - ";mo$"/"dy$"/"yr$: fl=1
2630 rem
2640 rem *** if clocks running update ut clock on screen ***
2650 rem
2660 if ft=0 then 2710
2670 t$=ti$: print "■■";tab(2);left$(t$,2)"▪|";mid$(t$,3,2);"▪|";right$(t$,2)
2680 rem
2690 rem *** if motors running check joystick correction ***
2700 rem
2710 if fs=0 then 3070
2720 t4=peek(r1+1): t1=peek(r1)+t4*256: t2=val(rm$): t3=val(rh$)
2730 if t1=0 then 2870
2740 if t4>127 then t1=65536-t1: t1=-t1
2750 if abs(t1)<sd then 2870
2760 if abs(t1)>=sd then t2=t2+sgn(t1): t1=t1-sd*sgn(t1): goto 2760
2770 if t1=0 then poke r1,0: poke r1+1,0: goto 2800
```

```
2780 if t1<0 then t1=abs(65536+t1)
2790 t3=int(t1/256): poke r1+1,t3: poke r1,t1-t3*256
2800 t3=val(rh$)
2810 if t2>59 then t2=t2-60: t3=t3+1: goto 2810
2820 if t2<0 then t2=t2+60: t3=t3-1: goto 2820
2830 if t3<0 then t3=t3+24: goto 2830
2840 if t3>23 then t3=t3-24: goto 2840
2850 rm$=str$(t2): rm$=right$(rm$,len(rm$)-1): if len(rm$)=1 then rm$="0"+rm$
2860 rh$=str$(t3): rh$=right$(rh$,len(rh$)-1): if len(rh$)=1 then rh$="0"+rh$
2870 t3=t1: t4=peek(d1+1): t1=peek(d1)+t4*256: t2=val(dc$)
2880 if t1=0 then if t3=0 then 3070
2890 if t4>127 then t1=65536-t1: t1=-t1
2900 if abs(t1)<.1*ds then 3050
2910 if abs(t1)>=.1*ds then t2=t2+.1*sgn(t1): t1=t1-.1*ds*sgn(t1): goto 2910
2920 if t1=0 then poke d1,0: poke d1+1,0: goto 2950
2930 if t1<0 then t1=abs(65536+t1)
2940 t3=int(t1/256): poke d1+1,t3: poke d1,t1-t3*256
2950 t3=0
2960 if t2>90 then t2=90-(t2-90): goto 2960
2970 if t2<-90 then t2=-180-t2: goto 2970
2980 t2=int(10*t2)/10: dc$=str$(t2)
2990 if int(val(dc$))=val(dc$) then dc$=dc$+".0"
3000 if left$(dc$,1)="-" then 3020
3010 dc$="+"+right$(dc$,len(dc$)-1)
3020 if val(dc$)=0 then dc$="+00.0": goto 3050
3030 if abs(val(dc$))<1 then dc$=left$(dc$,1)+"0"+right$(dc$,len(dc$)-1)
3040 if abs(val(dc$))<10 then dc$=left$(dc$,1)+"0"+right$(dc$,len(dc$)-1)
3050 print "■■■■■■";tab(10);"| r.a.   ";rh$;" h ";rm$;" m";tab(28);"|"
3060 print tab(10);"| dec.   ";dc$;" deg";tab(28);"|"
3070 if a>0 then if a<8 then print left$(cd$,2*a+8);tab(8);" "
3080 goto 1870
3090 rem
3100 rem *** keyboard input routine ***
3110 rem
3120 input#1,t$: print: return
3130 rem
3140 rem *** get a keystroke ***
3150 rem
3160 get a$: if a$<>"" then 3160
3170 get a$: if a$="" then 3170
3180 return
3190 rem
3200 rem *** poke time into start buffer ***
3210 rem
3220 for i=1 to 5 step 2: d=val(mid$(t$,i,1))
3230 d=d*16+val(mid$(t$,i+1,1))
3240 poke t+2-(i-1)/2,d: next i
3250 if left$(t$,2)="12" then k=128-k
3260 poke t+2, peek (t+2) and 127: poke t+2, peek(t+2) or k: return
3270 rem
3280 rem *** set telescope reference point ***
3290 rem
3300 sys ue
3310 gosub 3650
3320 print "■";chr$(142);"■point the telescope to an object with"
3330 print "with known coorinates. when ready to"
3340 print "begin tracking, press the space bar."
3350 gosub 3160: if a$<>" " then 3350
3360 sys en: sys sp: fs=1: print "■■"
3370 print "enter declination as +/-dd.d? ";
3380 gosub 3120: dc$=t$: if left$(dc$,1)="+" then 3400
3390 if left$(dc$,1)<>"-" then print "first character must be +/-.": goto 3370
3400 if len(dc$)<>5 then print "■declination must be +/-dd.d!": goto 3370
3410 if val(dc$)<-90 or val(dc$)>90 then print "■out of range!": goto 3370
3420 print "■■now enter the right ascension of the"
3430 print "object as hh,mm? ";
```

```
3440 input#1,rh$,rm$:.print
3450 if len(rh$)<>2 then print "■hour part must be 2 digits!": goto 3420
3460 t1=val(rh$)
3470 if t1<0 or t1>23 then print "■hours out of range!": goto 3420
3480 if len(rm$)<>2 then print "■minute part must be 2 digits!": goto 3420
3490 t1=val(rm$)
3500 if t1<0 or t1>59 then print "■minutes out of range!": goto 3420
3510 gosub 4030: t1=val(rh$)
3520 if t1>20 then if h<(25-w) then h=h+24: goto 3540
3530 if t1>16 then if e>4 then if t1>0 then t1=t1+e-4
3540 t1=abs(t1-h)
3550 if t1<=4 then return
3560 a$=str$(w): a$=right$(a$,len(a$)-1)
3570 if w<10 then a$="0"+a$
3580 t$=str$(e): t$=right$(t$,len(a$)-1)
3590 if e<10 then t$="0"+t$
3600 print "■valid r.a. range is ";a$;" hr to ";t$;" hr": goto 3420
3610 return
3620 rem
3630 rem *** turn off interrupts if on ***
3640 rem
3650 if fs=1 then sys cs: fs=0: rem disable stepper motor
3660 if ft=1 then sys cl: ft=0: rem disable time display
3670 return
3680 rem
3690 rem *** put clock display on the screen ***
3700 rem
3710 print "■UCuniversalCIUCstandardCCIUCCClocalCCCI"
3720 print "|    :   :      ||    :   :      ||    :   :      |"
3730 print "JCCCCCCCCCCKJCCCCCCCCCCCKJCCCCCCCCCCCK"
3740 print tab(11);"date - ";mo$"/"dy$"/"yr$
3750 print tab(10);"UCCCCCCCCCCCCCCCCCI"
3760 print tab(10);"| r.a.   ";rh$;" h ";rm$;" m";tab(28);"|"
3770 print tab(10);"| dec.   ";dc$;" deg";tab(28);"|"
3780 print tab(10);"JCCCCCCCCCCCCCCCCCK■"
3790 return
3800 rem
3810 rem *** display action menu ***
3820 rem
3830 print "■■■■■■■■■"
3840 print tab(6);"UCCCCCCCCCCCCCCCCCCCCCCCCI"
3850 print tab(6);"|    ■1■ reset reference      |"
3860 print tab(6);"|";spc(24);"|"
3870 print tab(6);"|    ■2■ slew telescope       |"
3880 print tab(6);"|";spc(24);"|"
3890 print tab(6);"|    ■3■ start tracking       |"
3900 print tab(6);"|";spc(24);"|"
3910 print tab(6);"|    ■4■ stop tracking        |"
3920 print tab(6);"|";spc(24);"|"
3930 print tab(6);"|    ■5■ stop clocks          |"
3940 print tab(6);"|";spc(24);"|"
3950 print tab(6);"|    ■6■ restart clocks       |"
3960 print tab(6);"|";spc(24);"|"
3970 print tab(6);"|    ■7■ exit program         |"
3980 print tab(6);"JCCCCCCCCCCCCCCCCCCCCCCCCK"
3990 return
4000 rem
4010 rem *** calculate local zenith ***
4020 rem
4030 t$=ti$: h=val(left$(t$,2)): m=val(mid$(t$,3,2))
4040 t1=dt*nd+gm+(((tz+h))/24)*dt: h=int(t1)
4050 t1=60*(t1-h): t1=int(t1)
4060 m=m+t1+val(mm$)*cr
4070 if m<0 then m=m+60: h=h-1: goto 4070
4080 if m>59 then m=m-60: h=h+1: goto 4080
4090 if h<0 then h=h+24: goto 4090
```

```
4100 if h>23 then h=h-24: goto 4100
4110 fg=0: e=h+4: if e>24 then e=e-24: fg=1
4120 w=h-4: if w<0 then w=w+24: fg=fg+99
4130 return

ready.
```

VARIABLE DESCRIPTIONS FOR THE BASIC PORTION OF THE TELESCOPE CONTROL SOFTWARE

A —Used as a flag and the menu selection variable.
A$ —Used as keyboard input string with the GET command.
B —AM/PM flag for local standard time.
C —AM/PM flag for local time.
CD$ —String of cursor down characters for formatting.
CL —Address of machine-language routine to disable time display interrupts.
CR —Flag for direction of local time correction.
CS —Address of machine-language routine to disable stepper motor interrupts.
D —Temporary day number in calendar calculations.
D(—Array with number of days plus one in each month.
D1 —Address of correction for joystick declination adjustments.
DC —Address for number of steps to slew in declination.
DC$ —Current value for declination of telescope.
DN —Direction byte for declination slew.
DS —Number of steps to move telescope one degree in declination.
DT —Daily change in Greenwich Mean Sidereal Time. Used to calculate local Zenith.
DY —Numeric value for day number.
DY$ —String for day number.
E —East limit for right ascension travel of telescope.
EN —Address of routine to energize stepper motors.
EP —Number of days from date 0/0/0 to 1/0/1985. Used to calculate local zenith.
FF —Flag variable.
FG —Flag variable.
FL —Flag variable.
FS —Flag to indicate whether stepper motor interrupts are active or not.
FT —Flag to indicate whether clock interrupts are active or not.
GM —Greenwich Mean Sidereal Time at 0 hr on 1/0/1985.
H —Hour part or right ascension of local zenith.
HH —Used in calculation of local time.
I —Loop counter.
K —Temporary for AM/PM flag of timers.
LN —Local longitude.

LP —Flag for result of leap year calculation.
LT —Local time storage buffer.
LT$ —String containing local time.
M —Temporary month number in calendar calculations.
MD —Number of steps to slew in declination.
MM —Used in calculation of local time.
MM$—Used in calculation of local time.
MO —Numeric value for month number.
MO$ —String for month number.
MR —Number of steps to slew in right ascension.
N —Hour part of destination right ascension for slew.
ND —Number of days since 1/0/1985.
R1 —Address of correction for joystick right ascension adjustments.
RA —Address for number of steps to slew in right ascension.
RH$ —Hour part of current value for right ascension.
RM$ —Minute part of current value for right ascension.
RN —Direction byte for right ascension slew.
RS —Address of routine to restart time display interrupts.
SD —Number of steps to move telescope 1 minute in right ascension.
SL —Address of routine to slew telescope to new location.
SP —Address of routine to set up stepper motor interrupts.
SS —Used in calculation of local time.
SS$ —Used in calculation of local time.
T —Temporary variable.
T$ —Used as keyboard input string with the INPUT command.
T1 —Temporary variable.
T2 —Temporary variable.
T3 —Temporary variable.
T4 —Temporary variable.
TC$ —Temporary copy of current declination.
TH$ —Temporary copy of hours part of current right ascension.
TI$ —Time variable in BASIC. Used for Universal time.
TM —Address of routine to initialize time display interrupts.
TM$ —Local Standard time string and temporary copy of minutes part of current right ascension.
TT —Address of local Standard time storage buffer.
TZ —Time zone for telescope location.
UE —Address of routine to put stepper motors in free wheel state.
UT —Used in calculation of universal time.
UT$ —Initial value of Universal time.
W —West limit for right ascension travel of telescope.
Y —Temporary year number in calendar calculations.
YR —Numeric value for year number.
YR$ —String for year number.

Listing A-2. BASIC Loader for Machine-Language Subroutines.

```
100 REM STEPPER CONTROL MACHINE LANGUAGE SUBROUTINES
110 REM WRITTEN BY RICHARD F. DALEY
120 REM FOR THE COMMODORE-64
130 REM
140 REM COPYRIGHT 1985
150 REM PERMISSION TO COPY
160 REM BUT NOT TO SELL OR DISTRIBUTE
170 REM
180 PRINT "{CLR}{DOWN}{DOWN}";TAB(11);"STEPPER SUBROUTINES"
190 PRINT TAB(16);"{DOWN}{DOWN}(C) 1985"
200 PRINT TAB(19);"{DOWN}{DOWN}BY"
210 PRINT TAB(12);"RICHARD F. DALEY"
220 PRINT "{DOWN}{DOWN}NOW SAVING THE STEPPER ROUTINES ON THE": PRINT "DISK. . . ."
230 PO=49152: HI=INT(PO/256): LO=PO-256*HI
240 READ A: IF A<>256 THEN CK=CK+A: GOTO 240
250 IF CK<>139068 THEN PRINT "ERROR IN DATA STATEMENTS!": END
260 RESTORE: OPEN 15,8,15
270 OPEN 2,8,3,"0:STEP.EX,P,W"
280 GOSUB 500
290 PRINT#2,CHR$(LO);: PRINT#2,CHR$(HI);
300 READ A: IF A=256 THEN 330
310 PRINT#2,CHR$(A);: GOSUB 500
320 GOTO 300
330 CLOSE 2
340 PRINT "{DOWN}{DOWN}PROGRAM STEP.EX SUCCESSFULLY SAVED."
350 PRINT "LOAD AND RUN THE CONTROL PROGRAM FOR"
360 PRINT "A CHECK.": END
500 INPUT#15,EN,ER$,TR,SC:IF EN=0 THEN RETURN
510 PRINT "{CLR}{DOWN}{DOWN}{DOWN}DISK ERROR -": PRINT EN;",";ER$;",";TR;",";SC
520 CLOSE 2
530 PRINT#15,"S0:STEP.EX"
540 CLOSE 15: END
1000 DATA 0, 0, 0, 0, 0, 0, 0, 0, 0, 0, 0, 0
1010 DATA 0, 0, 0, 0, 0, 0, 76, 62, 192, 76, 205, 192
1020 DATA 76, 233, 192, 76, 17, 195, 76, 47, 195, 76, 189, 195
1030 DATA 76, 102, 196, 76, 117, 196, 0, 0, 0, 0, 0, 0
1040 DATA 0, 0, 0, 0, 0, 0, 0, 0, 0, 0, 0, 0
1050 DATA 0, 0, 120, 173, 24, 3, 141, 159, 192, 173, 25, 3
1060 DATA 141, 160, 192, 169, 140, 141, 24, 3, 169, 192, 141, 25
1070 DATA 3, 169, 58, 141, 7, 221, 169, 33, 141, 6, 221, 169
1080 DATA 255, 141, 3, 221, 169, 0, 141, 53, 192, 169, 20, 141
1090 DATA 54, 192, 169, 0, 141, 1, 221, 169, 17, 141, 15, 221
1100 DATA 173, 13, 221, 169, 130, 141, 13, 221, 169, 1, 141, 43
1110 DATA 192, 169, 56, 141, 42, 192, 88, 96, 72, 138, 72, 152
1120 DATA 72, 172, 13, 221, 152, 41, 2, 208, 8, 104, 168, 104
1130 DATA 170, 104, 76, 0, 0, 206, 42, 192, 208, 8, 32, 183
1140 DATA 192, 169, 56, 141, 42, 192, 32, 204, 195, 104, 168, 104
1150 DATA 170, 104, 64, 173, 53, 192, 76, 192, 192, 173, 54, 192
1160 DATA 9, 8, 141, 1, 221, 234, 234, 169, 0, 141, 1, 221
1170 DATA 96, 120, 173, 159, 192, 141, 24, 3, 173, 160, 192, 141
1180 DATA 25, 3, 173, 13, 221, 169, 128, 141, 13, 221, 169, 0
1190 DATA 141, 15, 221, 88, 96, 32, 205, 192, 173, 9, 192, 16
1200 DATA 8, 173, 53, 192, 9, 6, 76, 254, 192, 173, 53, 192
1210 DATA 9, 4, 141, 53, 192, 32, 214, 194, 78, 49, 192, 110
1220 DATA 48, 192, 110, 47, 192, 78, 52, 192, 110, 51, 192, 110
1230 DATA 50, 192, 173, 49, 192, 13, 48, 192, 208, 4, 173, 47
1240 DATA 192, 44, 169, 255, 141, 55, 192, 141, 56, 192, 173, 52
1250 DATA 192, 13, 51, 192, 208, 4, 173, 50, 192, 44, 169, 255
1260 DATA 141, 57, 192, 141, 58, 192, 32, 214, 194, 173, 55, 192
1270 DATA 74, 74, 141, 61, 192, 56, 173, 47, 192, 237, 61, 192
1280 DATA 141, 47, 192, 173, 48, 192, 233, 0, 141, 48, 192, 173
1290 DATA 49, 192, 233, 0, 141, 49, 192, 173, 56, 192, 74, 74
1300 DATA 141, 61, 192, 56, 173, 47, 192, 237, 61, 192, 141, 47
1310 DATA 192, 173, 48, 192, 233, 0, 141, 48, 192, 173, 49, 192
```

```
1320 DATA 233, 0, 141, 49, 192, 173, 57, 192, 74, 74, 141, 61
1330 DATA 192, 56, 173, 50, 192, 237, 61, 192, 141, 50, 192, 173
1340 DATA 51, 192, 233, 0, 141, 51, 192, 173, 52, 192, 233, 0
1350 DATA 141, 52, 192, 173, 58, 192, 74, 74, 141, 61, 192, 56
1360 DATA 173, 50, 192, 237, 61, 192, 141, 50, 192, 173, 51, 192
1370 DATA 233, 0, 141, 51, 192, 173, 52, 192, 233, 0, 141, 52
1380 DATA 192, 169, 255, 141, 60, 192, 173, 55, 192, 240, 29, 173
1390 DATA 60, 192, 32, 237, 194, 32, 183, 192, 56, 173, 60, 192
1400 DATA 233, 4, 141, 60, 192, 56, 173, 55, 192, 233, 4, 141
1410 DATA 55, 192, 176, 222, 173, 47, 192, 13, 48, 192, 13, 49
1420 DATA 192, 208, 3, 76, 36, 194, 169, 0, 32, 237, 194, 32
1430 DATA 183, 192, 206, 47, 192, 173, 47, 192, 201, 255, 208, 224
1440 DATA 206, 48, 192, 173, 48, 192, 201, 255, 208, 214, 206, 49
1450 DATA 192, 173, 49, 192, 201, 255, 208, 204, 169, 0, 141, 60
1460 DATA 192, 173, 56, 192, 240, 29, 173, 60, 192, 32, 237, 194
1470 DATA 32, 183, 192, 24, 173, 60, 192, 105, 4, 141, 60, 192
1480 DATA 56, 173, 56, 192, 233, 4, 141, 56, 192, 176, 222, 169
1490 DATA 0, 141, 53, 192, 32, 62, 192, 169, 255, 141, 60, 192
1500 DATA 173, 57, 192, 240, 29, 173, 60, 192, 32, 237, 194, 32
1510 DATA 189, 192, 56, 173, 60, 192, 233, 4, 141, 60, 192, 56
1520 DATA 173, 57, 192, 233, 4, 141, 57, 192, 176, 222, 173, 50
1530 DATA 192, 13, 51, 192, 13, 52, 192, 208, 3, 76, 174, 194
1540 DATA 169, 0, 32, 237, 194, 32, 189, 192, 206, 50, 192, 173
1550 DATA 50, 192, 201, 255, 208, 224, 206, 51, 192, 173, 51, 192
1560 DATA 201, 255, 208, 214, 206, 52, 192, 173, 52, 192, 201, 255
1570 DATA 208, 204, 169, 0, 141, 60, 192, 173, 58, 192, 240, 29
1580 DATA 173, 60, 192, 32, 237, 194, 32, 189, 192, 24, 173, 60
1590 DATA 192, 105, 4, 141, 60, 192, 56, 173, 58, 192, 233, 4
1600 DATA 141, 58, 192, 176, 222, 96, 160, 2, 185, 6, 192, 153
1610 DATA 47, 192, 136, 16, 247, 160, 2, 185, 12, 192, 153, 50
1620 DATA 192, 136, 16, 247, 96, 208, 3, 76, 9, 195, 141, 59
1630 DATA 192, 162, 255, 32, 11, 195, 162, 255, 32, 11, 195, 162
1640 DATA 122, 32, 11, 195, 206, 59, 192, 208, 236, 162, 168, 202
1650 DATA 208, 253, 234, 234, 96, 162, 3, 189, 255, 191, 157, 8
1660 DATA 220, 202, 208, 247, 162, 3, 189, 2, 192, 157, 8, 221
1670 DATA 202, 208, 247, 169, 0, 141, 8, 220, 141, 8, 221, 173
1680 DATA 20, 3, 205, 187, 195, 240, 3, 141, 187, 195, 173, 21
1690 DATA 3, 205, 188, 195, 240, 3, 141, 188, 195, 120, 169, 82
1700 DATA 141, 20, 3, 169, 195, 141, 21, 3, 88, 96, 162, 3
1710 DATA 160, 0, 189, 8, 220, 41, 112, 74, 74, 74, 74, 24
1720 DATA 105, 48, 153, 55, 4, 200, 189, 8, 220, 41, 15, 24
1730 DATA 105, 48, 153, 55, 4, 200, 200, 202, 208, 224, 173, 11
1740 DATA 220, 48, 3, 169, 1, 44, 169, 16, 153, 55, 4, 173
1750 DATA 8, 220, 162, 3, 160, 0, 189, 8, 221, 41, 112, 74
1760 DATA 74, 74, 74, 24, 105, 48, 153, 68, 4, 200, 189, 8
1770 DATA 221, 41, 15, 24, 105, 48, 153, 68, 4, 200, 200, 202
1780 DATA 208, 224, 173, 11, 221, 48, 3, 169, 1, 44, 169, 16
1790 DATA 153, 68, 4, 173, 8, 221, 76, 0, 0, 120, 173, 187
1800 DATA 195, 141, 20, 3, 173, 188, 195, 141, 21, 3, 88, 96
1810 DATA 173, 43, 192, 73, 1, 141, 43, 192, 208, 3, 76, 101
1820 DATA 196, 173, 0, 220, 162, 0, 160, 0, 74, 176, 1, 136
1830 DATA 74, 176, 1, 200, 74, 176, 1, 202, 74, 176, 1, 232
1840 DATA 74, 176, 3, 169, 1, 44, 169, 0, 141, 46, 192, 142
1850 DATA 44, 192, 140, 45, 192, 173, 44, 192, 240, 45, 48, 14
1860 DATA 32, 183, 192, 238, 10, 192, 208, 35, 238, 11, 192, 76
1870 DATA 51, 196, 173, 53, 192, 72, 9, 2, 141, 53, 192, 32
1880 DATA 183, 192, 206, 10, 192, 173, 10, 192, 201, 255, 208, 3
1890 DATA 206, 11, 192, 104, 141, 53, 192, 173, 45, 192, 240, 45
1900 DATA 48, 14, 32, 189, 192, 238, 16, 192, 208, 35, 238, 17
1910 DATA 192, 76, 101, 196, 173, 54, 192, 72, 9, 2, 141, 54
1920 DATA 192, 32, 189, 192, 206, 16, 192, 173, 16, 192, 201, 255
1930 DATA 208, 3, 206, 17, 192, 104, 141, 54, 192, 96, 173, 1
1940 DATA 0, 141, 1, 221, 234, 234, 173, 17, 0, 141, 1, 221
1950 DATA 96, 169, 0, 141, 1, 221, 234, 234, 169, 16, 141, 1
1960 DATA 221, 96, 256

READY.
```

VARIABLE DESCRIPTIONS FOR THE BASIC LOADER FOR MACHINE LANGUAGE SUBROUTINES

A —Temporary variable for data read routine.
CK —Checksum variable.
EN —Error number from disk.
ER$ —Error message from disk.
HI —High byte of starting address.
LO —Low byte of starting address.
PO —Starting address for machine language subroutines.
SC —Sector number for disk error message.
TR —Track number for disk error message.

Listing A-3. Source Listing of Machine-Language Subroutines.

```
0010 ;       step.cont
0020 ;
0030 ;   ******************************************
0040 ;   *                                        *
0050 ;   *      Interrupt driven program to       *
0060 ;   *     drive two stepper motors using     *
0070 ;   *       the HS-3 controller board        *
0080 ;   *             produced by:               *
0090 ;   *                                        *
0100 ;   *             CyberPak                   *
0110 ;   *             P.O. Box 38                *
0120 ;   *             Brookfield, IL 60513       *
0130 ;   *             (312) 387-0802             *
0140 ;   *                                        *
0150 ;   *     This program written by:           *
0160 ;   *                                        *
0170 ;   *        Dr. Richard F. Daley            *
0180 ;   *        Copyright 1985                  *
0190 ;   *        All rights reserved.            *
0200 ;   *                                        *
0210 ;   *        Published by -                  *
0220 ;   *        Tab Books, Inc.                 *
0230 ;   *                                        *
0240 ;   ******************************************
0250 ;
0260 ;   *** Memory locations ***
0270 ;
0280 cinv       .de $0314           ;vector - IRQ svc routine
0290 nmivec     .de $0318           ;vector - NMI svc routine
0300 stdloc     .de $0437           ;screen for standard time
0310 locloc     .de $0444           ;screen for local time
0320 joystk     .de $dc00           ;joystick port 2
0330 stod       .de $dc08           ;tod for standard time
0340 cia2       .de $dd00           ;address of CIA2 registers
0350 portb      .de cia2+$01        ;user port latch
0360 ddr        .de cia2+$03        ;data direction register
0370 timb       .de cia2+$06        ;timer B latch
0380 ltod       .de cia2+$08        ;tod for local time
0390 icr        .de cia2+$0d        ;interrupt ctrl register
0400 crb        .de cia2+$0f        ;clock control register
0410 ;
0420 ;   *** Constants ***
0430 ;
0440 skip       .de $38             ;number of times to skip s
```

```
                 0450 step.delay .de $3a21           ;step interval for 1.2 pps
                 0460 ;
                 0470 ;    *** Start of program ***
                 0480 ;
                 0490            .ba $c000           ;start code at $c000
                 0500 ;          .os
                 0510 ;
                 0520 ;    *** Buffers for starting time ***
                 0530 ;
C000-            0540 stdtm      .ds $03             ;standard time start
C003-            0550 loctm      .ds $03             ;local time start
                 0560 ;
                 0570 ;    *** Buffers for slew data ***
                 0580 ;
C006-            0590 raslew     .ds $03             ;steps for slew in R.A. ax
C009-            0600 radir      .ds $01             ;slew direction in R.A.
C00A-            0610 ra.adj     .ds $02             ;joystick adjustment for R
C00C-            0620 decslew    .ds $03             ;steps for slew in Dec. ax
C00F-            0630 decdir     .ds $01             ;slew direction in Dec.
C010-            0640 dec.adj    .ds $02             ;joystick adjustment for D
                 0650 ;
                 0660 ;    *** Jump table for routines ***
                 0670 ;
C012- 4C 3E C0   0680            jmp start           ;enable interrupt routine
C015- 4C CD C0   0690            jmp clear           ;disable interrupt routine
C018- 4C E9 C0   0700            jmp slew            ;slew both r.a and dec
C01B- 4C 11 C3   0710            jmp time            ;setup TOD clocks
C01E- 4C 2F C3   0720            jmp time20          ;restart time display
C021- 4C BD C3   0730            jmp clrtime         ;clear time display
C024- 4C 66 C4   0740            jmp frewheel        ;set steppers to freewheel
C027- 4C 75 C4   0750            jmp energize        ;energize steppers
                 0760 ;
                 0770 ;    *** Internal storage locations ***
                 0780 ;
C02A-            0790 count      .ds $01             ;count to skip interrupt
C02B-            0800 dojoy      .ds $01             ;flag to skip joystick
C02C-            0810 dx         .ds $01             ;x status of joystick
C02D-            0820 dy         .ds $01             ;y status of joystick
C02E-            0830 fire       .ds $01             ;status of fire button
C02F-            0840 tempra     .ds $03             ;copy of slew for R.A.
C032-            0850 tempdec    .ds $03             ;copy of slew for Dec.
C035-            0860 xstepbuf   .ds $01             ;status of R.A. stepper
C036-            0870 ystepbuf   .ds $01             ;status of Dec. stepper
C037-            0880 ra.rampup  .ds $01             ;steps to ramp up in R.A.
C038-            0890 ra.rampdn  .ds $01             ;steps to ramp down in R.A
C039-            0900 dec.rampup .ds $01             ;steps to ramp up in Dec.
C03A-            0910 dec.rampdn .ds $01             ;steps to ramp down in Dec
C03B-            0920 dly.const  .ds $01             ;delay scaler value
C03C-            0930 ramp.ctr   .ds $01             ;counter for ramp up/down
C03D-            0940 temp       .ds $01             ;temporary storage
                 0950 ;
                 0960 ; Timer B of CIA2 is used to generate an NMI interrupt.
                 0970 ; The initial section of the code is used to set up the
                 0980 ; NMI vector at $0318 to point to the service routine
                 0990 ; intsvc. This routine will check to see if timer B
                 1000 ; generated the interrupt. If so then step the stepper
                 1010 ; motor. Otherwise transfer control to the NMI service
                 1020 ; routine originally vectored at nmivec.
                 1030 ;
                 1040 ;    *** Intercept NMI interrupts ***
                 1050 ;
C03E- 78         1060 start      sei                 ;disable interrupts
C03F- AD 18 03   1070            lda nmivec          ;save the current NMI
C042- 8D 9F C0   1080            sta donmi+1         ; service routine address
C045- AD 19 03   1090            lda nmivec+1        ; for jmp at domni
C048- 8D A0 C0   1100            sta donmi+2
C04B- A9 8C      1110            lda #l,intsvc       ;point nmivec to the
```

```
C04D- 8D 18 03  1120            sta nmivec          ; routine at intsvc
C050- A9 C0     1130            lda #h,intsvc
C052- 8D 19 03  1140            sta nmivec+1
C055- A9 3A     1150            lda #h,step.delay   ;setup delay for timer B
C057- 8D 07 DD  1160            sta timb+1
C05A- A9 21     1170            lda #l,step.delay
C05C- 8D 06 DD  1180            sta timb
C05F- A9 FF     1190            lda #$ff            ;set user port to output
C061- 8D 03 DD  1200            sta ddr
C064- A9 00     1210            lda #%00000000      ;set R.A. to half step
C066- 8D 35 C0  1220            sta xstepbuf        ; ccw motion
C069- A9 14     1230            lda #%00010100      ; and Dec. to full step
C06B- 8D 36 C0  1240            sta ystepbuf        ; ccw motion
C06E- A9 00     1250            lda #%00000000      ;set user port bits to
C070- 8D 01 DD  1260            sta portb           ; all outputs
C073- A9 11     1270            lda #%00010001      ;start timer B
C075- 8D 0F DD  1280            sta crb
C078- AD 0D DD  1290            lda icr             ;clear icr
C07B- A9 82     1300            lda #%10000010      ;set icr to interrupt on
C07D- 8D 0D DD  1310            sta icr             ; underflow from timer B
C080- A9 01     1320            lda #$01            ;enable joystick check
C082- 8D 2B C0  1330            sta dojoy
C085- A9 38     1340            lda #skip
C087- 8D 2A C0  1350            sta count
C08A- 58        1360            cli                 ;enable interrupts .
C08B- 60        1370            rts
                1380 ;
                1390 ;    *** Interrupt service routine ***
                1400 ;
C08C- 48        1410 intsvc     pha                 ;registers saved on stack
C08D- 8A        1420            txa
C08E- 48        1430            pha
C08F- 98        1440            tya
C090- 48        1450            pha
C091- AC 0D DD  1460            ldy icr             ;get the contents of icr
C094- 98        1470            tya
C095- 29 02     1480            and #%10            ;mask for timer B interrup
C097- D0 08     1490            bne svc.step        ;yes - step the motor
C099- 68        1500            pla                 ;no - restore the register
C09A- A8        1510            tay
C09B- 68        1520            pla
C09C- AA        1530            tax
C09D- 68        1540            pla
C09E- 4C 00 00  1550 donmi      jmp $0000           ;jump - os NMI svc routine
                1560 ;
C0A1- CE 2A C0  1570 svc.step   dec count
C0A4- D0 08     1580            bne svc.step10
C0A6- 20 B7 C0  1590            jsr step            ;step the motor
C0A9- A9 38     1600            lda #skip
C0AB- 8D 2A C0  1610            sta count
C0AE- 20 CC C3  1620 svc.step10 jsr chkjoy          ;check joystick status
C0B1- 68        1630            pla                 ;restore the registers
C0B2- A8        1640            tay
C0B3- 68        1650            pla
C0B4- AA        1660            tax
C0B5- 68        1670            pla
C0B6- 40        1680            rti                 ;return from the interrupt
                1690 ;
                1700 ;    *** One step to the R.A. stepper motor ***
                1710 ;
C0B7- AD 35 C0  1720 step       lda xstepbuf        ;get R.A. stepper cmd
C0BA- 4C C0 C0  1730            jmp dostep
                1740 ;
                1750 ;    *** One step to the Dec. stepper motor ***
                1760 ;
C0BD- AD 36 C0  1770 ystep      lda ystepbuf        ;get Dec. stepper cmd
C0C0- 09 08     1780 dostep     ora #%00001000      ;now move the stepper
```

```
C0C2- 8D 01 DD  1790            sta portb          ; one step
C0C5- EA        1800            nop
C0C6- EA        1810            nop
C0C7- A9 00     1820            lda #%00000000
C0C9- 8D 01 DD  1830            sta portb
C0CC- 60        1840            rts
                1850 ;
                1860 ;    *** Restore old NMI vector ***
                1870 ;
C0CD- 78        1880 clear      sei                ;disable interrupts
C0CE- AD 9F C0  1890            lda donmi+1        ;restore old NMI vector
C0D1- 8D 18 03  1900            sta nmivec
C0D4- AD A0 C0  1910            lda donmi+2
C0D7- 8D 19 03  1920            sta nmivec+1
C0DA- AD 0D DD  1930            lda icr
C0DD- A9 80     1940            lda #%10000000     ;clear interrupt mask
C0DF- 8D 0D DD  1950            sta icr
C0E2- A9 00     1960            lda #%00000000     ;disable timer B
C0E4- 8D 0F DD  1970            sta crb
C0E7- 58        1980            cli                ;enable interrupts
C0E8- 60        1990            rts
                2000 ;
                2010 ;    *** Slew the stepper to new location ***
                2020 ;
C0E9- 20 CD C0  2030 slew       jsr clear          ;clear stepper interrupt
C0EC- AD 09 C0  2040            lda radir          ;set R.A. stepper directio
C0EF- 10 08     2050            bpl slew1
C0F1- AD 35 C0  2060            lda xstepbuf
C0F4- 09 06     2070            ora #%00000110     ;set R.A. to full step
C0F6- 4C FE C0  2080            jmp slew2          ; and cw direction
                2090 ;
C0F9- AD 35 C0  2100 slew1      lda xstepbuf       ;get port status
C0FC- 09 04     2110            ora #%00000100     ;set R.A. to full step
C0FE- 8D 35 C0  2120 slew2      sta xstepbuf
C101- 20 D6 C2  2130            jsr move
C104- 4E 31 C0  2140            lsr tempra+2       ;divide R.A. by 2
C107- 6E 30 C0  2150            ror tempra+1
C10A- 6E 2F C0  2160            ror tempra
C10D- 4E 34 C0  2170            lsr tempdec+2      ;divide Dec. by 2
C110- 6E 33 C0  2180            ror tempdec+1
C113- 6E 32 C0  2190            ror tempdec
C116- AD 31 C0  2200            lda tempra+2       ;check if only ramp
C119- 0D 30 C0  2210            ora tempra+1       ; needed
C11C- D0 04     2220            bne ramp           ;ramp only
C11E- AD 2F C0  2230            lda tempra         ;set steps to ramp
C121- 2C        2240            .by $2c
C122- A9 FF     2250 ramp       lda #$ff
C124- 8D 37 C0  2260            sta ra.rampup      ;equal number up
C127- 8D 38 C0  2270            sta ra.rampdn      ; and down
C12A- AD 34 C0  2280            lda tempdec+2      ;check if only ramp
C12D- 0D 33 C0  2290            ora tempdec+1      ; needed
C130- D0 04     2300            bne d.ramp         ;ramp only
C132- AD 32 C0  2310            lda tempdec        ;set steps to ramp
C135- 2C        2320            .by $2c
C136- A9 FF     2330 d.ramp     lda #$ff
C138- 8D 39 C0  2340            sta dec.rampup     ;equal number up
C13B- 8D 3A C0  2350            sta dec.rampdn     ; and down
C13E- 20 D6 C2  2360            jsr move           ;get slew in work area
C141- AD 37 C0  2370            lda ra.rampup      ;subtract ramp up from sle
C144- 4A        2380            lsr a              ;divide by 4
C145- 4A        2390            lsr a
C146- 8D 3D C0  2400            sta temp
C149- 38        2410            sec                ;subtract ramp up from sle
C14A- AD 2F C0  2420            lda tempra
C14D- ED 3D C0  2430            sbc temp
C150- 8D 2F C0  2440            sta tempra
```

```
C153- AD 30 C0   2450            lda tempra+1
C156- E9 00      2460            sbc #$00
C158- 8D 30 C0   2470            sta tempra+1
C15B- AD 31 C0   2480            lda tempra+2
C15E- E9 00      2490            sbc #$00
C160- 8D 31 C0   2500            sta tempra+2
C163- AD 38 C0   2510            lda ra.rampdn         ;subtract ramp down from s
C166- 4A         2520            lsr a                 ;divide by 4
C167- 4A         2530            lsr a
C168- 8D 3D C0   2540            sta temp
C16B- 38         2550            sec
C16C- AD 2F C0   2560            lda tempra
C16F- ED 3D C0   2570            sbc temp
C172- 8D 2F C0   2580            sta tempra
C175- AD 30 C0   2590            lda tempra+1
C178- E9 00      2600            sbc #$00
C17A- 8D 30 C0   2610            sta tempra+1
C17D- AD 31 C0   2620            lda tempra+2
C180- E9 00      2630            sbc #$00
C182- 8D 31 C0   2640            sta tempra+2
C185- AD 39 C0   2650            lda dec.rampup        ;subtract ramp up from sle
C188- 4A         2660            lsr a                 ;divide by 4
C189- 4A         2670            lsr a
C18A- 8D 3D C0   2680            sta temp
C18D- 38         2690            sec
C18E- AD 32 C0   2700            lda tempdec
C191- ED 3D C0   2710            sbc temp
C194- 8D 32 C0   2720            sta tempdec
C197- AD 33 C0   2730            lda tempdec+1
C19A- E9 00      2740            sbc #$00
C19C- 8D 33 C0   2750            sta tempdec+1
C19F- AD 34 C0   2760            lda tempdec+2
C1A2- E9 00      2770            sbc #$00
C1A4- 8D 34 C0   2780            sta tempdec+2
C1A7- AD 3A C0   2790            lda dec.rampdn        ;subtract ramp down from s
C1AA- 4A         2800            lsr a                 ;divide by 4
C1AB- 4A         2810            lsr a
C1AC- 8D 3D C0   2820            sta temp
C1AF- 38         2830            sec
C1B0- AD 32 C0   2840            lda tempdec
C1B3- ED 3D C0   2850            sbc temp
C1B6- 8D 32 C0   2860            sta tempdec
C1B9- AD 33 C0   2870            lda tempdec+1
C1BC- E9 00      2880            sbc #$00
C1BE- 8D 33 C0   2890            sta tempdec+1
C1C1- AD 34 C0   2900            lda tempdec+2
C1C4- E9 00      2910            sbc #$00
C1C6- 8D 34 C0   2920            sta tempdec+2
C1C9- A9 FF      2930            lda #$ff
C1CB- 8D 3C C0   2940            sta ramp.ctr
C1CE- AD 37 C0   2950 slew10     lda ra.rampup         ;ramp R.A. motor up
C1D1- F0 1D      2960            beq slew20
C1D3- AD 3C C0   2970            lda ramp.ctr
C1D6- 20 ED C2   2980            jsr delay
C1D9- 20 B7 C0   2990            jsr step
C1DC- 38         3000            sec                   ;decrement by 4
C1DD- AD 3C C0   3010            lda ramp.ctr
C1E0- E9 04      3020            sbc #$04
C1E2- 8D 3C C0   3030            sta ramp.ctr
C1E5- 38         3040            sec
C1E6- AD 37 C0   3050            lda ra.rampup
C1E9- E9 04      3060            sbc #$04
C1EB- 8D 37 C0   3070            sta ra.rampup
C1EE- B0 DE      3080            bcs slew10            ;go again if >=0
C1F0- AD 2F C0   3090 slew20     lda tempra            ;finished main slew yet?
C1F3- 0D 30 C0   3100            ora tempra+1
C1F6- 0D 31 C0   3110            ora tempra+2
```

```
C1F9- D0 03      3120              bne slew30          ; no do next step
C1FB- 4C 24 C2   3130              jmp slew40          ; yes ramp down
                 3140 ;
C1FE- A9 00      3150 slew30       lda #$00
C200- 20 ED C2   3160              jsr delay
C203- 20 B7 C0   3170              jsr step
C206- CE 2F C0   3180              dec tempra          ;decrement slew count
C209- AD 2F C0   3190              lda tempra
C20C- C9 FF      3200              cmp #$ff
C20E- D0 E0      3210              bne slew20
C210- CE 30 C0   3220              dec tempra+1
C213- AD 30 C0   3230              lda tempra+1
C216- C9 FF      3240              cmp #$ff
C218- D0 D6      3250              bne slew20
C21A- CE 31 C0   3260              dec tempra+2
C21D- AD 31 C0   3270              lda tempra+2
C220- C9 FF      3280              cmp #$ff
C222- D0 CC      3290              bne slew20
C224- A9 00      3300 slew40       lda #$00            ;set ramp counter
C226- 8D 3C C0   3310              sta ramp.ctr
C229- AD 38 C0   3320 slew50       lda ra.rampdn       ;ramp R.A. motor down
C22C- F0 1D      3330              beq slew60
C22E- AD 3C C0   3340              lda ramp.ctr
C231- 20 ED C2   3350              jsr delay
C234- 20 B7 C0   3360              jsr step
C237- 18         3370              clc                 ;add 4 to counter
C238- AD 3C C0   3380              lda ramp.ctr
C23B- 69 04      3390              adc #$04
C23D- 8D 3C C0   3400              sta ramp.ctr
C240- 38         3410              sec                 ;subtract 4 from ramp down
C241- AD 38 C0   3420              lda ra.rampdn
C244- E9 04      3430              sbc #$04
C246- 8D 38 C0   3440              sta ra.rampdn
C249- B0 DE      3450              bcs slew50          ;again if rampdn >=0
C24B- A9 00      3460 slew60       lda #%00000000      ;restore half step and ccw
C24D- 8D 35 C0   3470              sta xstepbuf        ; R.A. stepper
C250- 20 3E C0   3480              jsr start           ;restart stepper interrupt
C253- A9 FF      3490              lda #$ff
C255- 8D 3C C0   3500              sta ramp.ctr
C258- AD 39 C0   3510 dslew10      lda dec.rampup      ;ramp Dec. motor up
C25B- F0 1D      3520              beq dslew20
C25D- AD 3C C0   3530              lda ramp.ctr
C260- 20 ED C2   3540              jsr delay
C263- 20 BD C0   3550              jsr ystep
C266- 38         3560              sec                 ;decrement by 4
C267- AD 3C C0   3570              lda ramp.ctr
C26A- E9 04      3580              sbc #$04
C26C- 8D 3C C0   3590              sta ramp.ctr
C26F- 38         3600              sec
C270- AD 39 C0   3610              lda dec.rampup
C273- E9 04      3620              sbc #$04
C275- 8D 39 C0   3630              sta dec.rampup
C278- B0 DE      3640              bcs dslew10         ;go again if >=0
C27A- AD 32 C0   3650 dslew20      lda tempdec         ;finished main slew yet?
C27D- 0D 33 C0   3660              ora tempdec+1
C280- 0D 34 C0   3670              ora tempdec+2
C283- D0 03      3680              bne dslew30         ; no do next step
C285- 4C AE C2   3690              jmp dslew40         ; yes ramp down .
                 3700 ;
C288- A9 00      3710 dslew30      lda #$00
C28A- 20 ED C2   3720              jsr delay
C28D- 20 BD C0   3730              jsr ystep
C290- CE 32 C0   3740              dec tempdec         ;decrement slew count
C293- AD 32 C0   3750              lda tempdec
C296- C9 FF      3760              cmp #$ff
C298- D0 E0      3770              bne dslew20
```

```
C29A- CE 33 C0  3780               dec tempdec+1
C29D- AD 33 C0  3790               lda tempdec+1
C2A0- C9 FF     3800               cmp #$ff
C2A2- D0 D6     3810               bne dslew20
C2A4- CE 34 C0  3820               dec tempdec+2
C2A7- AD 34 C0  3830               lda tempdec+2
C2AA- C9 FF     3840               cmp #$ff
C2AC- D0 CC     3850               bne dslew20
C2AE- A9 00     3860 dslew40       lda #$00           ;set ramp counter
C2B0- 8D 3C C0  3870               sta ramp.ctr
C2B3- AD 3A C0  3880 dslew50       lda dec.rampdn     ;ramp Dec. motor down
C2B6- F0 1D     3890               beq dslew60
C2B8- AD 3C C0  3900               lda ramp.ctr
C2BB- 20 ED C2  3910               jsr delay
C2BE- 20 BD C0  3920               jsr ystep
C2C1- 18        3930               clc                ;add 4 to counter
C2C2- AD 3C C0  3940               lda ramp.ctr
C2C5- 69 04     3950               adc #$04
C2C7- 8D 3C C0  3960               sta ramp.ctr
C2CA- 38        3970               sec                ;subtract 4 from ramp down
C2CB- AD 3A C0  3980               lda dec.rampdn
C2CE- E9 04     3990               sbc #$04
C2D0- 8D 3A C0  4000               sta dec.rampdn
C2D3- B0 DE     4010               bcs dslew50        ;again if rampdn >=0
C2D5- 60        4020 dslew60       rts                ;slew completed!
                4030 ;
                4040 ;   *** Copy slew values to work area ***
                4050 ;
C2D6- A0 02     4060 move          ldy #$02           ;copy slew in R.A. to
C2D8- B9 06 C0  4070 move10        lda raslew,y       ; temp area
C2DB- 99 2F C0  4080               sta tempra,y
C2DE- 88        4090               dey
C2DF- 10 F7     4100               bpl move10
C2E1- A0 02     4110               ldy #$02           ;copy slew in Dec. to
C2E3- B9 0C C0  4120 move20        lda decslew,y      ; temp area
C2E6- 99 32 C0  4130               sta tempdec,y
C2E9- 88        4140               dey
C2EA- 10 F7     4150               bpl move20
C2EC- 60        4160               rts
                4170 ;
                4180 ;   *** Delay subroutine ***
                4190 ;
                4200 ; on entry delay 852 cycles if .a=0
                4210 ; and delay 833,427 cycles if .a=255
                4220 ;
C2ED- D0 03     4230 delay         bne delay10        ;if .a<>0 then calc delay
C2EF- 4C 09 C3  4240               jmp delay1000
                4250 ;
C2F2- 8D 3B C0  4260 delay10       sta dly.const      ;save constant
C2F5- A2 FF     4270 delay500      ldx #$ff           ;delay 3247 cycles
C2F7- 20 0B C3  4280               jsr delay.lp       ; per count in dly.const
C2FA- A2 FF     4290               ldx #$ff
C2FC- 20 0B C3  4300               jsr delay.lp
C2FF- A2 7A     4310               ldx #$7a
C301- 20 0B C3  4320               jsr delay.lp
C304- CE 3B C0  4330               dec dly.const      ;see if delay finished
C307- D0 EC     4340               bne delay500       ; if <>0 then not done
C309- A2 A8     4350 delay1000     ldx #$a8           ;delay 852 cycles
C30B- CA        4360 delay.lp      dex                ; 5 cycles per loop
C30C- D0 FD     4370               bne delay.lp
C30E- EA        4380               nop                ;plus 10 cycles at end
C30F- EA        4390               nop
C310- 60        4400               rts
                4410 ;
                4420 ;   *** Set up TOD clocks and interrupts ***
                4430 ;
C311- A2 03     4440 time          ldx #$03           ;initialize standard TOD
```

```
C313- BD FF BF  4450             lda stdtm-1,x         ; clock
C316- 9D 08 DC  4460             sta stod,x
C319- CA        4470             dex
C31A- D0 F7     4480             bne time+2
C31C- A2 03     4490 time10      ldx #$03              ;initialize local TOD
C31E- BD 02 C0  4500             lda loctm-1,x         ; clock
C321- 9D 08 DD  4510             sta ltod,x
C324- CA        4520             dex
C325- D0 F7     4530             bne time10+2
C327- A9 00     4540             lda #$00              ;set tenths second clock
C329- 8D 08 DC  4550             sta stod              ; to zero and start clocks
C32C- 8D 08 DD  4560             sta ltod
C32F- AD 14 03  4570 time20      lda cinv              ;save cinv vector
C332- CD BB C3  4580             cmp timext+1
C335- F0 03     4590             beq time30
C337- 8D BB C3  4600             sta timext+1
C33A- AD 15 03  4610 time30      lda cinv+1
C33D- CD BC C3  4620             cmp timext+2
C340- F0 03     4630             beq time40
C342- 8D BC C3  4640             sta timext+2
C345- 78        4650 time40      sei                   ;reset cinv vector
C346- A9 52     4660             lda #l,timedisp
C348- 8D 14 03  4670             sta cinv
C34B- A9 C3     4680             lda #h,timedisp
C34D- 8D 15 03  4690             sta cinv+1
C350- 58        4700             cli
C351- 60        4710             rts
                4720 ;
                4730 ;    *** Display current reading in TOD clocks ***
                4740 ;
C352- A2 03     4750 timedisp    ldx #$03
C354- A0 00     4760             ldy #$00
C356- BD 08 DC  4770 prtdigit    lda stod,x            ;get TOD register
C359- 29 70     4780             and #$70              ;now get high digit
C35B- 4A        4790             lsr a                 ;move to low nybble
C35C- 4A        4800             lsr a
C35D- 4A        4810             lsr a
C35E- 4A        4820             lsr a
C35F- 18        4830             clc                   ;convert to ASCII
C360- 69 30     4840             adc #'0
C362- 99 37 04  4850             sta stdloc,y          ;display it
C365- C8        4860             iny
C366- BD 08 DC  4870             lda stod,x
C369- 29 0F     4880             and #$0f              ;get low digit
C36B- 18        4890             clc                   ;convert to ASCII
C36C- 69 30     4900             adc #'0
C36E- 99 37 04  4910             sta stdloc,y          ;display it
C371- C8        4920             iny
C372- C8        4930             iny
C373- CA        4940             dex
C374- D0 E0     4950             bne prtdigit          ;do the next one
C376- AD 0B DC  4960             lda stod+$03
C379- 30 03     4970             bmi timed10
C37B- A9 01     4980             lda #$01              ;screen for a
C37D- 2C        4990             .by $2c
C37E- A9 10     5000 timed10     lda #$10              ;screen for p
C380- 99 37 04  5010             sta stdloc,y
C383- AD 08 DC  5020             lda stod              ;reset latched time
C386- A2 03     5030             ldx #$03              ;now do local time
C388- A0 00     5040             ldy #$00
C38A- BD 08 DD  5050 lprtdigit   lda ltod,x            ;get TOD register
C38D- 29 70     5060             and #$70              ;now get high digit
C38F- 4A        5070             lsr a                 ;move to low nybble
C390- 4A        5080             lsr a
C391- 4A        5090             lsr a
C392- 4A        5100             lsr a
```

```
C393- 18          5110          clc                   ;convert to ASCII
C394- 69 30       5120          adc #'0
C396- 99 44 04    5130          sta locloc,y          ;display it
C399- C8          5140          iny
C39A- BD 08 DD    5150          lda ltod,x
C39D- 29 0F       5160          and #$0f              ;get low digit
C39F- 18          5170          clc                   ;convert to ASCII
C3A0- 69 30       5180          adc #'0
C3A2- 99 44 04    5190          sta locloc,y          ;display it
C3A5- C8          5200          iny
C3A6- C8          5210          iny
C3A7- CA          5220          dex
C3A8- D0 E0       5230          bne lprtdigit         ;do the next one
C3AA- AD 0B DD    5240          lda ltod+$03
C3AD- 30 03       5250          bmi timed20
C3AF- A9 01       5260          lda #$01              ;screen for a
C3B1- 2C          5270          .by $2c
C3B2- A9 10       5280 timed20  lda #$10              ;screen for p
C3B4- 99 44 04    5290          sta locloc,y
C3B7- AD 08 DD    5300          lda ltod
C3BA- 4C 00 00    5310 timext   jmp $0000             ;jump to do rest of irq
                  5320 ;
                  5330 ;   *** Disconnect this IRQ service routine ***
                  5340 ;
C3BD- 78          5350 clrtime  sei                   ;disable interrupts
C3BE- AD BB C3    5360          lda timext+1          ;restore old IRQ vector
C3C1- 8D 14 03    5370          sta cinv
C3C4- AD BC C3    5380          lda timext+2
C3C7- 8D 15 03    5390          sta cinv+1
C3CA- 58          5400          cli                   ;enable interrupts
C3CB- 60          5410          rts
                  5420 ;
                  5430 ;   *** Check joystick and move stepper ***
                  5440 ;
C3CC- AD 2B C0    5450 chkjoy   lda dojoy             ;check only on alternate
C3CF- 49 01       5460          eor #$01              ; calls to routine
C3D1- 8D 2B C0    5470          sta dojoy
C3D4- D0 03       5480          bne chkjoy5
C3D6- 4C 65 C4    5490          jmp extjoy
                  5500 ;
C3D9- AD 00 DC    5510 chkjoy5  lda joystk            ;get data from port
C3DC- A2 00       5520          ldx #$00
C3DE- A0 00       5530          ldy #$00
C3E0- 4A          5540          lsr a                 ;put lo bit in carry
C3E1- B0 01       5550          bcs chkjoy10          ;carry set - skip
C3E3- 88          5560          dey                   ;joystick pointed up
C3E4- 4A          5570 chkjoy10 lsr a                 ;next bit in carry
C3E5- B0 01       5580          bcs chkjoy20          ;skip
C3E7- C8          5590          iny                   ;joystick pointed down
C3E8- 4A          5600 chkjoy20 lsr a                 ;get next bit
C3E9- B0 01       5610          bcs chkjoy30          ;skip if set
C3EB- CA          5620          dex                   ;joystick pointed left
C3EC- 4A          5630 chkjoy30 lsr a                 ;get next bit
C3ED- B0 01       5640          bcs chkjoy40          ;skip
C3EF- E8          5650          inx                   ;joystick pointed right
C3F0- 4A          5660 chkjoy40 lsr a                 ;check fire button
C3F1- B0 03       5670          bcs chkjoy50
C3F3- A9 01       5680          lda #$01              ;fire pressed
C3F5- 2C          5690          .by $2c
C3F6- A9 00       5700 chkjoy50 lda #$00              ;fire not pressed
C3F8- 8D 2E C0    5710          sta fire
C3FB- 8E 2C C0    5720          stx dx                ;save joystick registers
C3FE- 8C 2D C0    5730          sty dy
C401- AD 2C C0    5740          lda dx                ;check x register
C404- F0 2D       5750          beq do.ystep          ;no change in y
C406- 30 0E       5760          bmi movx              ;move stepper back
C408- 20 B7 C0    5770 doxstep  jsr step              ;do step in x direction
```

```
C40B- EE 0A C0  5780             inc ra.adj
C40E- D0 23     5790             bne do.ystep
C410- EE 0B C0  5800             inc ra.adj+1
C413- 4C 33 C4  5810             jmp do.ystep
                5820 ;
C416- AD 35 C0  5830 movx        lda xstepbuf        ;change stepper direction
C419- 48        5840             pha                 ;save stepper command word
C41A- 09 02     5850             ora #%10
C41C- 8D 35 C0  5860             sta xstepbuf
C41F- 20 B7 C0  5870 doxstep10   jsr step            ;now move stepper
C422- CE 0A C0  5880             dec ra.adj
C425- AD 0A C0  5890             lda ra.adj
C428- C9 FF     5900             cmp #$ff
C42A- D0 03     5910             bne doxstep20
C42C- CE 0B C0  5920             dec ra.adj+1
C42F- 68        5930 doxstep20   pla                 ;restore normal stepper
C430- 8D 35 C0  5940             sta xstepbuf        ; command word
C433- AD 2D C0  5950 do.ystep    lda dy              ;check y register
C436- F0 2D     5960             beq extjoy          ;no movement
C438- 30 0E     5970             bmi movy            ;move y higher
C43A- 20 BD C0  5980 doystep     jsr ystep           ;step Dec. motor
C43D- EE 10 C0  5990             inc dec.adj
C440- D0 23     6000             bne extjoy
C442- EE 11 C0  6010             inc dec.adj+1
C445- 4C 65 C4  6020             jmp extjoy
                6030 ;
C448- AD 36 C0  6040 movy        lda ystepbuf        ;setup to move y lower
C44B- 48        6050             pha
C44C- 09 02     6060             ora #%10
C44E- 8D 36 C0  6070             sta ystepbuf
C451- 20 BD C0  6080 doystep10   jsr ystep           ;move Dec. motor
C454- CE 10 C0  6090             dec dec.adj
C457- AD 10 C0  6100             lda dec.adj
C45A- C9 FF     6110             cmp #$ff
C45C- D0 03     6120             bne doystep20
C45E- CE 11 C0  6130             dec dec.adj+1
C461- 68        6140 doystep20   pla
C462- 8D 36 C0  6150             sta ystepbuf
C465- 60        6160 extjoy      rts
                6170 ;
                6180 ;    *** Set steppers to freewheel ***
                6190 ;
C466- AD 01 00  6200 frewheel    lda %00000001       ;set R.A. stepper to
C469- 8D 01 DD  6210             sta portb           ; to freewheel state
C46C- EA        6220             nop
C46D- EA        6230             nop
C46E- AD 11 00  6240             lda %00010001       ;set Dec. stepper to
C471- 8D 01 DD  6250             sta portb           ; freewheel state
C474- 60        6260             rts
                6270 ;
                6280 ;    *** Energize steppers ***
                6290 ;
C475- A9 00     6300 energize    lda #%00000000      ;set R.A. stepper to
C477- 8D 01 DD  6310             sta portb           ; energize state
C47A- EA        6320             nop
C47B- EA        6330             nop
C47C- A9 10     6340             lda #%00010000      ;set Dec. stepper to
C47E- 8D 01 DD  6350             sta portb           ; energize state
C481- 60        6360             rts
                6370 ;
                6380             .en
END OF MAE PASS!


 ---  LABEL FILE:  ---
```

```
chkjoy =C3CC          chkjoy10 =C3E4        chkjoy20 =C3E8
chkjoy30 =C3EC        chkjoy40 =C3F0        chkjoy5 =C3D9
chkjoy50 =C3F6        cia2 =DD00            cinv =0314
clear =C0CD           clrtime =C3BD         count =C02A
crb =DD0F             d.ramp =C136          ddr =DD03
dec.adj =C010         dec.rampdn =C03A      dec.rampup =C039
decdir =C00F          decslew =C00C         delay =C2ED
delay.lp =C30B        delay10 =C2F2         delay1000 =C309
delay500 =C2F5        dly.const =C03B       do.ystep =C433
dojoy =C02B           donmi =C09E           dostep =C0C0
doxstep =C408         doxstep10 =C41F       doxstep20 =C42F
doystep =C43A         doystep10 =C451       doystep20 =C461
dslew10 =C258         dslew20 =C27A         dslew30 =C288
dslew40 =C2AE         dslew50 =C2B3         dslew60 =C2D5
dx =C02C              dy =C02D              energize =C475
extjoy =C465          fire =C02E            frewheel =C466
icr =DD0D             intsvc =C08C          joystk =DC00
locloc =0444          loctm =C003           lprtdigit =C38A
ltod =DD08            move =C2D6            move10 =C2D8
move20 =C2E3          movx =C416            movy =C448
nmivec =0318          portb =DD01           prtdigit =C356
ra.adj =C00A          ra.rampdn =C038       ra.rampup =C037
radir =C009           ramp =C122            ramp.ctr =C03C
raslew =C006          skip =0038            slew =C0E9
slew1 =C0F9           slew10 =C1CE          slew2 =C0FE
slew20 =C1F0          slew30 =C1FE          slew40 =C224
slew50 =C229          slew60 =C24B          start =C03E
stdloc =0437          stdtm =C000           step =C0B7
step.delay =3A21      stod =DC08            svc.step =C0A1
svc.step10 =C0AE      temp =C03D            tempdec =C032
tempra =C02F          timb =DD06            time =C311
time10 =C31C          time20 =C32F          time30 =C33A
time40 =C345          timed10 =C37E         timed20 =C3B2
timedisp =C352        timext =C3BA          xstepbuf =C035
ystep =C0BD           ystepbuf =C036
//0000,C482,C482
```

Appendix B

The Brightest Stars and Their Coordinates

FOLLOWING IS A LISTING OF THE BRIGHTEST STARS IN THE sky visible from approximately 40 degrees north latitude. Each of these listings includes the star's name, magnitude, constellation, right ascension, declination and month of the year when it is highest in the sky. See Table B-1.

While there are only 11 stars on the list, they provide good coverage for all of the year. The months listed are those when the star is high in the sky at 10 p.m. local time. Note that each of the stars are highest in the sky near the middle of the range of months listed. Near the beginning of the first month and the end of the last month listed, the star may be too low to be of use as a reference star.

Table B-1. The Brightest Stars.

Star	Mag. visual	Constellation	R.A. Hr Min (Epoch	Dec. deg. 1980.0)	Visible months
Aldebaran	0.86	Taurus	04 35	+16.5	1,11−12
Rigel	0.15	Orion	05 13	−08.2	1−2,12
Sirius	−1.42	Canis Major	06 44	−06.7	1-3
Procyon	0.35	Canis Minor	07 38	+05.3	2-4
Regulus	1.35	Leo	10 07	+12.1	3-6
Spica	0.96	Virgo	13 24	−11.0	4-6
Arcturus	−0.06	Bootes	14 14	+19.3	4-7
Antares	0.89	Scorpius	16 28	−26.4	6-9
Altair	0.74	Aquila	19 50	+08.8	7-9
Vega	0.03	Lyra	18 36	+38.8	7-10
Deneb	1.25	Cygnus	20 36	+45.2	7-11

Appendix C

Suppliers of Telescope Making Supplies

FOLLOWING IS A LIST OF SUPPLIERS THAT WE ARE FAMILIAR with. The comments about each supplier indicate what, if any, experience we have had with, or what knowledge we have gained about, that company. There are other suppliers that also may be reputable, so do not consider this list exhaustive.

1. **Coulter Optical Co.**
   P.O. Box K
   Idyllwild, CA 92349
   While we did not obtain any materials from Coulter Optical Co. for this project, we did purchase a mirror from them for a previous telescope project. The mirror they produced was a quality mirror and was delivered within two weeks of the promised date. From what we hear from other amateur-telescope makers, this also has been their experience.

2. **CyberPak Co.**
   P.O. Box 38
   Brookfield, IL 60513
   The stepper motors and interface board for this project were obtained from CyberPak Co. Also, they were very helpful to us as we worked to get this project up and running.

3. **Edmund Scientific**
   5290 Edscorp Bldg.
   Barrington, NJ 08007

   Edmund is one of the older companies in the field. However, they primarily make and sell completed telescopes for the mass market. We have a telescope from Edmund and while it is very good optically, the mount vibrates easily and does not stop quickly after changing the position of the telescope. Check out the catalog, as they do have a few things of interest to the ATM.

4. **Lumicon**
   2111 Research Dr. #5
   Livermore, CA 94550

   The focuser used in the construction of this telescope came from Lumicon. As well as focusers, Lumicon has a large supply of other useful items for the amateur astronomer. The president, Jack Marling, is an avid amateur astronomer. His interest is reflected in the items produced by the company—they are designed by someone who uses them.

5. **Meade Instruments Corporation**
   1675 Toronto Way
   Costa Mesa, CA 92626

   We have a Meade telescope and have been very happy with it. As well as selling completed telescopes, they also sell most of the items used in making their own telescopes separately. Check them out.

6. **Parks Optical**
   270 Easy Street
   Simi Valley, CA 93065

   Parks Optical is a reputable firm, but as we have had no dealings with them, we can make no personal observations about them. They have been around for a number of years making items for amateur-telescope makers which are said to be of high quality.

7. **Telescopics**
   P.O. Box 98
   La Canada, CA 91011

   The mirror for the telescope described in this book was obtained from this company. We were disturbed that there was no commitment to any delivery date, and, despite repeated calls, never could obtain any delivery date from them. Finally, we had to obtain another mirror so we could meet our publishing schedule. We waited in excess of 10 months with no indication of a delivery date before delivery was made. If time is a consideration, try to obtain a firm commitment from any telescope mirror maker before spending your money. On the other side of the story, the quality of their optics are well worth waiting for. *The mirror we finally received was outstanding.*

8. **TeleVue**
   20-A Dexter Plaza
   Pearl River, NY 10965

   TeleVue is best known for their eyepieces. They produced our favorite eyepiece, a 13 mm Nagler. They are good people to work with.

9. **Roger W. Tuthill, Inc.**
   11 Tanglewood Lane
   Mountainside, NJ 07092

   Several years ago we purchased a solar filter from Roger W. Tuthill, Inc. They delivered promptly, and we have been quite happy with the filter. Most of the items produced by them are more for the Schmidt-Cassegrain telescopes, but there are several items of interest to the amateur-telescope maker. Roger Tuthill is an avid amateur astronomer and his catalog shows it.

10. **University Optics**
    P.O. Box 1205
    Ann Arbor, MI 48106

    These people produce quality products and deliver them quickly. While we did not use any of their items in the construction of this telescope, we do have a number of their eyepieces and use them regularly when observing.

11. **Willmann-Bell, Inc.**
    P.O. Box 3125
    Richmond, VA 23235

    The catalog from Willmann-Bell, Inc. is about the most complete list of books about astronomy we have seen. We have ordered books from them a number of times, and they were courteous and shipped the merchandise promptly. They also have a line of supplies for amateur-telescope makers, but we have not used any of them nor have we heard much about them from other amateurs.

Appendix D

Suggested Items for Further Reading

THIS IS A LIST OF SOME BOOKS AND PERIODICALS THAT CONtain items of interest for further reading if you are interested in doing more in telescope construction. At the time of this writing, there seems to be only a few really useful resources for the amateur-telescope maker.

Also included are items related to astronomical computing—both books and software. There are not a large number of items available here either.

The comments made about each title represent our personal impressions of each one. We are only listing those titles already in our library or provided to us by the publishers for review. Please carefully evaluate, where possible, these items before making a purchase.

BOOKS

1. Berry, *How to Build a Dobsonian Telescope*, 2nd edition, AstroMedia Corp., Milwaukee, WI, 1982. Probably the most comprehensive set of instructions about the construction of a Dobsonian telescope. It is a compilation of articles from *Telescope Making* magazine on the topic.

2. Berry, *How to Build an Observatory*, AstroMedia Corp., Mil-

waukee, WI, 1981. This pamphlet is a compilation of articles from *Telescope Making* magazine on the topic of building your own observatory.

3. Brown, *All About Telescopes*, Edmund Scientific Co., Barrington, NJ, 1967. This book was written to feature various Edmund Scientific products, but the author has the ability to clearly and simply describe many ideas about telescopes and optics. That, along with the clear drawings, make this an ideal introductory book on the subject.

4. Burgess, *Celestial Basic*, Sybex, Berkeley, CA, 1982. This book is a compilation of useful software for astronomy. The programs in this book were written for the Apple II computer, but unfortunately few of them use many of the detailed features of the computer.

5. Genet, *Microcomputers in Astronomy*, Books I and II, Fairborn Observatory, Fairborn, OH, 1983, 1984. These books contain a series of papers on the use of computers in astronomy that were given in a symposium at Fairborn Observatory. These papers have many good ideas, but seldom contain enough detail to allow you to exactly duplicate the item described.

6. Ingalls, *Amateur Telescope Making*, Books I, II, and III, Scientific American, New York, 1935, 1937, 1953. This set of books is from a column appearing in *Scientific American* from the early 1920s to the mid-1950s. While many of the techniques and materials described are outdated, the design and technical items are pertinent to any telescope maker.

7. Trueblood and Genet, *Microcomputer Control of Telescopes*, Willmann-Bell, Inc., Richmond, VA, 1985. This book covers all of the details of computerizing your telescope. Unfortunately, many of the ideas and approaches they use are quite expensive, but they do clearly present the details of computerization.

8. Texereau, *How to Make a Telescope*, 2nd edition, Willmann-Bell, Inc., Richmond, VA, 1984. This book is the definitive work on telescope making, especially making your own optics. If you are serious about telescope construction, buy this book.

PERIODICALS

1. Gleanings for ATMs Column, *Sky and Telescope* magazine, Sky

Publishing Corp., Cambridge, MA. This monthly column contains many ideas from amateur-telescope makers around the world. Most of the time they require that you have some knowledge about making telescopes. Often this column is worth the price of the magazine; the rest is a nice bonus.

2. *Telescope Making* Magazine, AstroMedia Corp., Milwaukee, WI. This is a quarterly magazine devoted to telescope design and construction. If you have an interest in further telescope making, by all means subscribe.

SOFTWARE

1. Covitz, *Sky Travel*, Commodore Business Machines, West Chester, PA. This program is a well-written planetarium program. Impressive graphics and capabilities. Unfortunately, at this writing it is rumored that Commodore is no longer producing any software. Hopefully, another firm will offer this software.

2. Eagle, *Science Software*, available from David Eagle, 7952 W. Quarto Dr., Littleton, CO 80123. There are a number of programs available. We reviewed and liked the *Comet* program which predicts the position of a comet. We did not review other programs.

3. Kluepfel, *Astronomy Programs*, available from Charles Kluepfel, 11 George St., Bloomfield, NJ. There are a number of programs available. We particularly liked the programs *Moon & Sun, Eclipse Map* and *Lunar Eclipse Map*. These three work together and produce high resolution maps of the visibility of an eclipse of either the sun or the moon. Like these programs, the other programs seem well done and very useful, but slow.

4. *Indoor Astronomy*, Astro Link, P.O. Box 1978, Spring Valley, CA 92077. This program is a set of digitized images of astronomical objects. Unfortunately, we were provided with only a demo which consists of a set of pretty pictures. If these pictures are representative of the program, this can give you a lot of entertainment for observing. We cannot report on how the program works.

5. *Locator*, Interesting Software, P.O. Box 41-5012, Miami Beach, FL 33141. This program will update the position of a celestial object from one epoch to another and convert the coordinates to azimuth and elevation. This is especially useful for telescopes

with an altazimuth mount, such as the popular Dobsonian mount. The program seems to work well. Unfortunately, this program and manual are filled with distracting spelling and typographical errors. It is also very similar to programs commonly found in the public domain.

Appendix E

Observing Form

IT IS A GOOD PRACTICE TO MAKE NOTES AND DRAWINGS AT the eyepiece for all observing runs. Figure E-1 represents an observing form. This form may be photocopied, punched, and inserted into a notebook for future reference. Feel free to make changes to suit your needs.

Good sketches make you a more careful observer. While sketching an object, you will see more and more details as you progress through your sketch. Discoveries, such as a new comet or a supernova in a distant galaxy, must be carefully sketched or photographed when reporting to the proper agencies.

Meaningful observations are made by an observer who keeps a good set of records. When at the telescope, do not trust your memory to hold what your eye has seen. Record the information, when you see it. Every observer keeps records in a personalized fashion, but you should record the data shown on the following observing form.

Deep Sky Observing Form

Observation number ____________________

Catalog number ________________________ Date ___/___/___

Observing conditions (1-10) ___________ Time(UT) ____:____

Location ______________________________ Elevation ______ft

Latitude _____o _____' _____"

Longitude _____o _____' _____"

Observer ______________________________ Scope ____" f/____

Eyepiece 1 _____ mm Type _____ Power x_____ Filter _______

Eyepiece 2 _____ mm Type _____ Power x_____ Filter _______

NOTES:

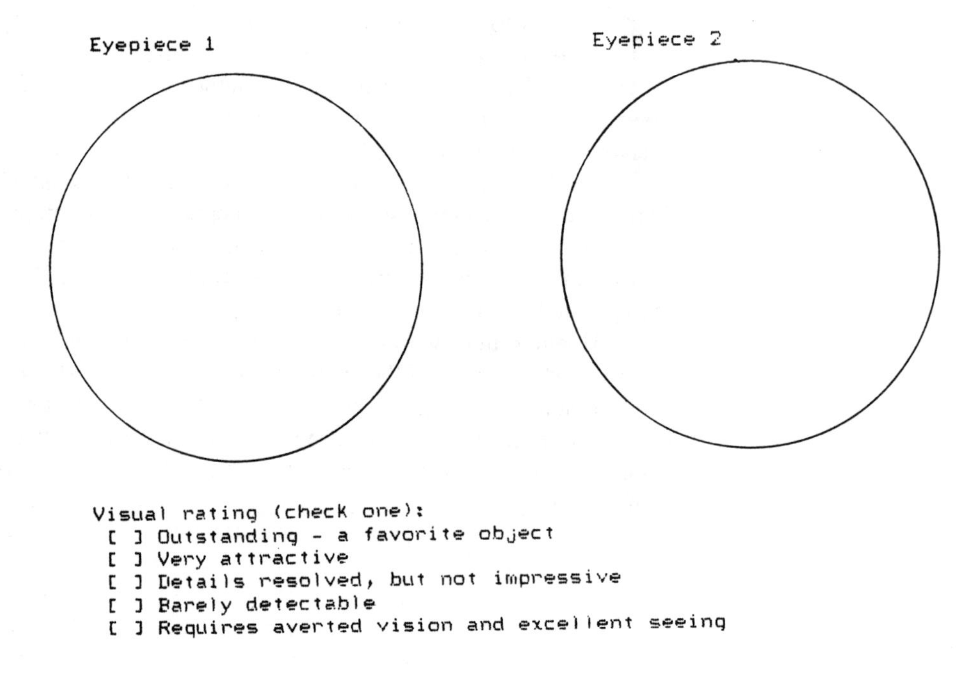

Eyepiece 1

Eyepiece 2

Visual rating (check one):

[ ] Outstanding - a favorite object
[ ] Very attractive
[ ] Details resolved, but not impressive
[ ] Barely detectable
[ ] Requires averted vision and excellent seeing

Fig.E-1.

Appendix F

An Astronomical Calendar Program

THERE ARE A NUMBER OF GENERAL ASTRONOMICAL DATA which are useful for you, an amateur astronomer, to know when planning the day and time of an observing session. Because the bright light from a full moon will interfere with you being able to see faint deep-sky objects, it is necessary to know the phase of the moon for a particular date. Also helpful in planning when to begin and end a particular observing session is the time the sun and moon rise and set. These facts, along with other data that might be of general interest to an amateur astronomer, are available with this astronomical calendar.

This appendix contains a listing of the astronomical calendar along with the necessary directions to use it. If you purchase the diskette containing the software to run your telescope from Easy-Ware, this program will also be on that diskette; otherwise type in Listing 4 as is found on the following pages. When you have finished entering it, save the program on a diskette using the name ASTROCAL. While this is a long program, it will produce a calendar containing a number of useful items.

The program will calculate various types of data about the sun and moon for any date you select after 1 A.D. These calculations are generally the most accurate between the dates 1900 and 2100 A.D. The further you go from these dates the less accurate will be the calculations—this is especially so for those dates involving the moon. The formulas used for calculating the rising and setting of the moon, as well as for calculating the moon's position in the sky, are in an abbreviated form. If the full equations for this data were used, the program would be more than twice as long as it presently is and would run much more slowly.

The sunrise/sunset times and the solstice and equinox times are quite accurate—usually within 3 minutes for any date within 300 years of the present. This should be quite accurate as—because of atmospheric conditions—most observers cannot pinpoint the rising or setting of the sun more accurately than that. Another factor that you must consider with both the sunrise/sunset and the moonrise/moonset times is your altitude. The nearer you are to sea level, with no obstructions of the horizon, the nearer to the actual time of rising or setting will be the visible rising or setting times. The moonrise/moonset times average about 8-10 minutes in error from the ephemeris tables. This means that they may be in error by 25-30 minutes on some days, while other days are quite close to the correct time.

These inaccuracies mean that this calendar is primarily useful for planning observing sessions and should not take the place of an ephemeris for your guide when studying lunar phenomenon. If, like the authors, you are most interested in observing deep sky objects, it is helpful to know the approximate rising and setting times for the sun and the moon as well as the moon's phases so you can plan your observing times accordingly.

Remember too, that the computer personalizes the astronomical calendar for your location. This may mean that it will not always agree with a general calendar which you may have. For example, the dates for the phases of the moon or equinox or solstice may vary from one day to another between your general calendar and this astronomical calendar, especially so when the time falls close to midnight.

If your location uses it, this calendar program will take into account whether daylight savings time is in effect as it calculates the times for each item of data. Because the program calculates Daylight Savings time in a simple manner, the calendar may not show the proper time adjustments for the last few days in April after Daylight Savings time goes into effect or for the last few days of October after Standard time is restored.

The program sets aside a space on the diskette for a file that will contain the information needed to personalize your calendar. Each time you load the calendar program, the computer will use this information. The first time you LOAD and RUN the program, the computer will look for this information and since the file is not there yet, the program will then ask you several questions. The first question is:

PLEASE ENTER YOUR NAME?

Entering your name will allow the computer to print your name on your calendar printout. Next the computer will ask:

WHAT IS YOUR LOCATION?

Again this is more information needed so that your calendar can be personalized. After you have entered your city name, you will be asked several questions to aid the computer in its calculations of the data specific for your area. You are first asked:

WHAT IS THE LATITUDE
OF your city?

The words "your city" will be replaced with the name you entered in the previous question. Please enter the latitude as accurately as possible, using a value accurate to at least a tenth of a degree. For locations in the southern hemisphere, enter the latitude as a negative value. This will be the same value as the one used in the control program. Next the computer will ask:

WHAT IS THE LONGITUDE
OF your city?

Enter your longitude, again the number here should be to the nearest tenth of a degree. The accuracy of the calculations for your location will depend on the accuracy of the latitude and longitude that you use. The next question is:

IN WHICH TIME ZONE IS
your city (0-23)?

You must enter a number for your time zone, as the computer will not recognize letters here. For the United States the time zones are Eastern, 5, Central, 6, Mountain, 7, and Pacific, 8. Finally, the computer will ask:

IS DAYLIGHT SAVINGS TIME USED IN
your city?

This time there is no cursor after the question mark. The question is to be answered with a yes or no answer, so press Y or N. For most areas in the United States the answer is yes.

Now you will have a chance to check all the data you have entered and make any corrections that you feel are necessary. The computer will ask:

IS THIS CORRECT?

Again, this is a yes or no question. If you want to leave the data as you have entered it, press the Y key. The computer will then save the data onto the diskette into the space mentioned earlier as being reserved for this file. If you did not enter the data just as you wanted it or if you want to look at what you entered to make sure it is correct, press the N key. The computer will then display each question along with the answer you had entered. If the information is correct, press the RETURN key, and the next question will be displayed. If the information is not correct, then edit your answers and press RETURN. Be careful not to press the CLR/HOME key. Use only the CURSOR CONTROL keys and the

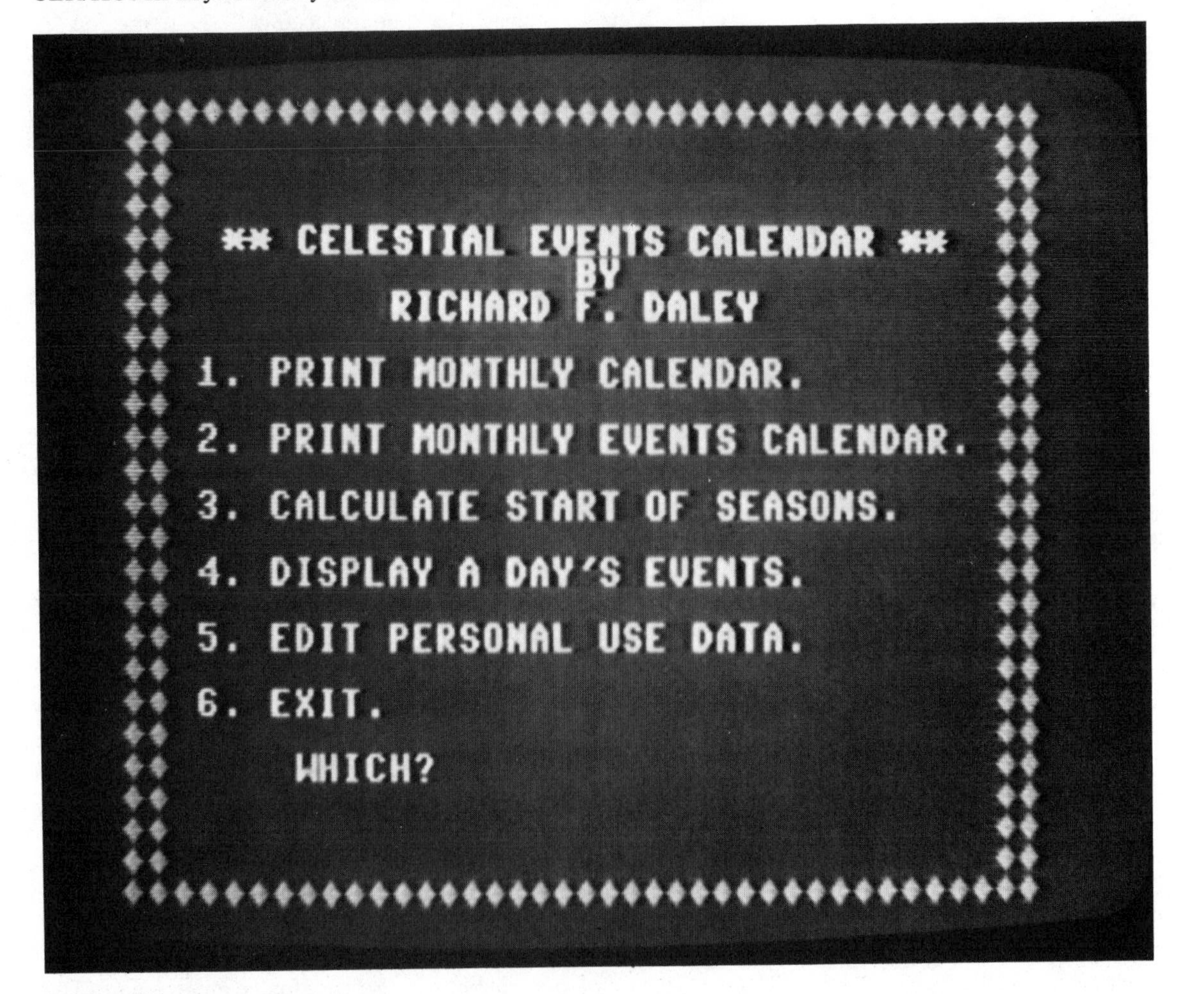

Fig. F-1. Main menus screen.

space bar to move the cursor about through your answer.

When you have entered all the information just the way you want it and have pressed the Y key to answer the IS THIS CORRECT? question, the screen will change colors and the menu of program options will appear as in Fig. F-1.

Listing F-1. Celestial Events Calendar Program.

```
100 REM  CELESTIAL EVENTS CALENDAR PROGRAM
110 REM  WRITTEN BY RICHARD F. DALEY
120 REM  FOR THE COMMODORE-64
130 REM
140 REM  COPYRIGHT 1986
150 REM  ILLEGAL TO COPY OR SELL
160 REM  OR DISTRIBUTE WITHOUT PERMISSION
170 REM
180 REM
190 REM
200 REM *** MONTH AND DAYS/MONTH ***
210 REM
220 DIM MO$(12),MO(12)
230 DATA JANUARY,31,FEBRUARY,28,MARCH,31,APRIL,30,MAY,31,JUNE,30,JULY,31
240 DATA AUGUST,31,SEPTEMBER,30,OCTOBER,31,NOVEMBER,30,DECEMBER,31
250 FOR I=1 TO 12: READ MO$(I),MO(I): NEXT I: OPEN 1,0,1: OPEN 15,8,15
260 REM
270 REM *** DAYS OF THE WEEK ***
280 REM
290 DIM DY$(7)
300 DATA SUNDAY,MONDAY,TUESDAY,WEDNESDAY,THURSDAY,FRIDAY,SATURDAY
310 FOR I=1 TO 7: READ DY$(I): NEXT I
320 REM
330 REM *** DATA FOR SEASON CALCULATIONS ***
340 REM
350 DIM LO(3),W$(3)
360 DATA SPRING,SUMMER,AUTUMN,WINTER
370 FOR I=0 TO 3: READ W$(I): NEXT I
380 FOR I=0 TO 3: LO(I)=I*90: NEXT I
390 REM
400 REM *** HOLIDAY INFORMATION ***
410 REM
420 REM     THE FORMAT IS (SO YOU CAN ADD YOUR OWN)
430 REM     FIRST NUMBER = MONTH OF HOLIDAY
440 REM     SECOND NUMBER = DATE OF HOLIDAY
450 REM           IF ZERO THEN FLOATING HOLIDAY
460 REM
470 DATA 1,1,NEW YEAR'S DAY
480 DATA 2,12,LINCOLN'S BIRTHDAY
490 DATA 2,14,VALENTINE'S DAY
500 DATA 2,0,WASHINGTON'S HOLIDAY
510 DATA 3,17,ST PATRICK'S DAY
520 DATA 5,0,MOTHER'S DAY
530 DATA 5,0,MEMORIAL DAY
540 DATA 6,14,FLAG DAY
550 DATA 6,0,FATHER'S DAY
560 DATA 7,4,INDEPENDANCE DAY
570 DATA 9,0,LABOR DAY
580 DATA 10,0,COLUMBUS DAY
590 DATA 10,31,HALOWEEN
600 DATA 11,0,ELECTION DAY
610 DATA 11,11,VETERAN'S DAY
620 DATA 11,0,THANKSGIVING
630 DATA 12,25,CHRISTMAS
640 HH=17: DIM HM(HH),HD(HH),HD$(HH)
```

```
650 FOR I=1 TO HH: READ HM(I),HD(I),HD$(I): NEXT I
660 REM
670 REM *** FINISH INITIALIZATION OF VARIABLES ***
680 REM
690 DIM EV$(31,4),P$(32),T(32),FF(7)
700 BL$="                              ": BL$=BL$+BL$
710 CL$="▌▌▌▌▌▌▌▌▌▌▌▌▌▌▌▌▌▌▌▌▌▌▌▌▌▌▌▌▌▌▌▌▌▌▌▌▌▌▌▌": CL$=CL$+CL$: POKE 808,234
720 CR$="▐▐▐▐▐▐▐▐▐▐▐▐▐▐▐▐▐▐▐▐▐▐▐▐▐▐▐▐▐▐▐▐▐▐▐▐▐▐▐▐": CR$=CR$+CR$
730 CD$="▒QQQQQQQQQQQQQQQQQQQQQQQQQQQQQQQQQQQQQQQ": NO$="■▒NO▒▒": YS$="■▒YES▒▒"
740 EP=715875: TP=π*2: RD=π/180: MR=29.530588
750 GOSUB 5180: IF A1$<>"00" AND TR=1 THEN 750
760 IF A1$<>"00" AND TR=0 THEN 6160
770 LA=VAL(LA$): LO=VAL(LO$)
780 REM
790 REM *** DEFINED FUNCTIONS ***
800 REM
810 DEF FN AC(X)=-ATN(X/SQR(-X*X+1))+TP/4
820 DEF FN RA(X)=X*RD
830 DEF FN DG(X)=X/RD
840 DEF FN SN(X)=SIN(X*RD)
850 DEF FN CS(X)=COS(X*RD)
860 DEF FN NM(X)=X-360*INT(X/360)
870 REM
880 REM *** MAIN MENU ENTRY ***
890 REM
900 POKE 53280,9: POKE 53281,9: PRINT "▒ ";CHR$(8);CHR$(142)
910 PRINT "▒": FOR I=0 TO 38: PRINT "▒";TAB(I);"◆";
920 PRINT LEFT$(CD$,25);TAB(38-I);"◆";: NEXT I
930 FOR I=0 TO 23
940 PRINT LEFT$(CD$,24-I);"◆◆";LEFT$(CD$,I+1);SPC(37);"◆◆": NEXT I
950 PRINT "▒";TAB(4);"▒QQQQQ** CELESTIAL EVENTS CALENDAR **"
960 PRINT TAB(19);"BY"
970 PRINT TAB(11);"RICHARD F. DALEY"
980 PRINT TAB(3);"▒▒1. PRINT MONTHLY CALENDAR."
990 PRINT TAB(3);"▒2. PRINT MONTHLY EVENTS CALENDAR."
1000 PRINT TAB(3);"▒3. CALCULATE START OF SEASONS."
1010 PRINT TAB(3);"▒4. DISPLAY A DAY'S EVENTS."
1020 PRINT TAB(3);"▒5. EDIT PERSONAL USE DATA."
1030 PRINT TAB(3);"▒6. EXIT."
1040 PRINT TAB(7);"▒WHICH?"
1050 GOSUB 1130: A=VAL(Q$): IF A<1 OR A>6 THEN 1050
1060 POKE 53280,6: POKE 53281,14: PRINT "▒"
1070 ON A GOTO 3960, 3920, 2550, 5720, 5260, 6160
1080 GOTO 1050
1090 REM
1100 REM *** GET A SINGLE KEYSTROKE ***
1110 REM
1120 PRINT "PRESS ANY KEY TO CONTINUE."
1130 GET Q$: IF Q$<>"" THEN 1130
1140 GET Q$: IF Q$="" THEN 1140
1150 RETURN
1160 REM
1170 REM *** CHECK VALIDITY OF A DATE ***
1180 REM
1190 ER=0: T$="MONTH": IF MO<1 OR MO>12 THEN 1260
1200 IF INT(MO)<>MO THEN 1280
1210 T$="DAY": IF DY<1 OR DY>MO(MO) THEN 1260
1220 IF INT(DY)<>DY THEN 1280
1230 IF YR<1 THEN 1290
1240 IF INT(YR)<>YR THEN T$="YEAR": GOTO 1280
1250 GOTO 1300
1260 IF DY>MO(MO) AND MO=2 THEN 1300
1270 PRINT "▒";T$;" OUT OF RANGE.": ER=1: RETURN
1280 PRINT "▒";T$;" MUST BE AN INTEGER.": ER=1: RETURN
1290 PRINT "▒YEAR BEFORE 1 A.D.": ER=1: RETURN
1300 YY=YR: IF MO>2 THEN 1360
```

```
1310 MO=MO+12: YR=YR-1
1320 LP=-(INT(YY/4)=(YY/4))+(INT(YY/100)=(YY/100))-(INT(YY/400)=(YY/400))
1330 T$="DAY": IF LP=1 AND VAL(MO$)=2 AND DY>29 THEN 1270
1340 IF LP=0 AND VAL(MO$)=2 AND DY=29 THEN 1430
1350 IF VAL(MO$)=2 AND DY>29 THEN 1270
1360 N=DY+2*MO+INT(.6*(MO+1))+YR+INT(YR/4)-INT(YR/100)+INT(YR/400)+2
1370 N=INT((N/7-INT(N/7))*7+.5)
1380 D5=1+2*MO+INT(.6*(MO+1))+YR+INT(YR/4)-INT(YR/100)+INT(YR/400)+2
1390 D5=INT((D5/7-INT(D5/7))*7+.5): IF N=0 THEN N=7
1400 IF D5<1 THEN D5=D5+7
1410 IF MO>12 THEN MO=MO-12
1420 RETURN
1430 PRINT "■IT IS NOT A LEAP YEAR": ER=1: RETURN
1440 REM
1450 REM *** FIND HOLIDAYS ***
1460 REM
1470 YY=YR: GOSUB 1940: IF ME=0 AND ED=0 THEN 1510
1480 IF ME=MO AND ED=DY THEN 1500
1490 GOTO 1510
1500 PRINT "■■THAT DAY IS EASTER.": GOSUB 1120: GOTO 900
1510 FOR I=1 TO HH: MM=HM(I): DD=HD(I): HD$=HD$(I)
1520 IF MM=MO THEN 1550
1530 IF MM>MO THEN I=HH
1540 NEXT I: RETURN
1550 IF DD=0 THEN JJ=0: GOSUB 1610: GOTO 1540
1560 IF DD=VAL(DY$) THEN JJ=0: GOSUB 1890
1570 GOTO 1540
1580 REM
1590 REM *** FIND FLOATING HOLIDAYS ***
1600 REM
1610 ON MM GOTO 1540, 1630, 1540, 1540, 1660, 1710
1620 ON MM-6 GOTO 1540, 1540, 1730, 1760, 1780, 1540
1630 T=D5: IF 21-D5+3>21 THEN T=D5+7
1640 T$=STR$(21-T+3): IF DY=VAL(T$) THEN 1890
1650 RETURN
1660 T$=STR$(14-D5+2): IF DY=14-D5+2 AND LEFT$(HD$,3)="MOT" THEN 1880
1670 IF LEFT$(HD$,3)="MOT" THEN RETURN
1680 T=31-D5: IF T<25 THEN T=T+7
1690 T$=STR$(T): IF DY=VAL(T$) AND LEFT$(HD$,3)="MEM" THEN 1880
1700 RETURN
1710 T$=STR$(21-D5+2): IF DY=21-D5+2 AND LEFT$(HD$,3)="FAT" THEN 1890
1720 RETURN
1730 T=10-D5: IF T>7 THEN T=T-7
1740 T$=STR$(T): IF DY=VAL(T$) THEN 1880
1750 RETURN
1760 T=17-D5: IF T>14 THEN T=T-7
1770 T$=STR$(T): IF DY=VAL(T$) THEN 1880
1780 RETURN
1790 T$="": IF LEFT$(HD$,1)="T" THEN 1850
1800 IF VAL(YR$)/2<>INT(VAL(YR$)/2) THEN RETURN
1810 T=11-D5: IF T>8 THEN T=T-7
1820 T$=STR$(T): IF DY=VAL(T$) THEN 1890
1830 RETURN
1840 GOTO 1880
1850 T$=STR$(20+(7-D5)): IF VAL(T$)<22 THEN T$=STR$(VAL(T$)+7)
1860 IF DY=VAL(T$) THEN 1890
1870 RETURN
1880 IF JJ=1 THEN RETURN
1890 JJ=2: PRINT"■■THAT DAY IS "HD$"."
1900 RETURN
1910 REM
1920 REM *** CALCULATE DATE OF EASTER - VALID FOR 1900 TO 2099 ***
1930 REM
1940 ED=0: ME=0: N=YY-1900: IF N<0 OR N>199 THEN RETURN
1950 M=VAL(MO$): IF M<3 OR M>4 THEN RETURN
1960 A=N/19: A=19*(A-INT(A)): B=INT((7*A+1)/19)
```

```
1970 M=(11*A+4.00001-B)/29: X=M-INT(M): IF X=1 THEN M=0: GOTO 1990
1980 IF X<>1 THEN M=29*X
1990 Q=INT(N/4): W=(N+Q+31-M)/7: W=INT(7*(W-INT(W))): DE=INT(25-M-W): D=DE
2000 IF DE<0 THEN ME=3
2010 IF DE>0 THEN ME=4
2020 IF DE=0 THEN ME=3: ED=31: RETURN
2030 IF DE<-9 THEN DE=DE+9: GOTO 2030
2040 IF DE<0 THEN D=31-ABS(DE)
2050 IF DE>0 THEN D=DE
2060 IF YY=1974 OR YY=1984 THEN D=D+7
2070 IF YY=1994 THEN D=D+7-31: ME=4
2080 ED=D: RETURN
2090 REM
2100 REM *** CALCULATE SIDEREAL TIME ***
2110 REM
2120 GC=6.61759: TC=.065711: TM=TC*YD+GC
2130 TT=TM: MA=24: GOSUB 2240: TM=TT
2140 RETURN
2150 REM
2160 REM *** CALCULATE RISE & SET TIMES FROM R.A AND DEC. ***
2170 REM
2180 TA=TAN(FN RA(LA))*TAN(FN RA(DM))
2190 TA=-TA
2200 TA=FN AC(TA): TA=FN DG(TA)
2210 TR=(RM-TA/15-TM)*.99726852: TT=TR: MA=24: GOSUB 2240: TR=TT
2220 TS=(RM+TA/15-TM)*.99726852: TT=TS: MA=24: GOSUB 2240: TS=TT
2230 RETURN
2240 TT=(TT/MA-INT(TT/MA))*MA: RETURN
2250 REM
2260 REM *** CALCULATE MOONRISE AND MOONSET TIMES ***
2270 REM
2280 M=VAL(MO$): D=VAL(DY$): Y=VAL(YR$)
2290 DG=0: IF M=1 THEN 2330
2300 FOR I=1 TO M-1: DG=DG+MO(I): NEXT I: IF M>2 THEN DG=DG+LP
2310 IF M=2 AND D=29 THEN DG=DG+1
2320 L1=LO: IF LO>180 THEN L1=-(LO-180)
2330 YD=DG+D+(L1/15)/24
2340 DG=YD+Y*365+INT(Y/4)-INT(Y/100)+INT(Y/400)
2350 NI=DG-EP: GOSUB 2120
2360 LZ=308.1687: LE=180.739: PL=253.7433
2370 PG=.110968*NI+PL: TT=PG: MA=360: GOSUB 2240: PG=TT
2380 LD=LZ+360*NI/27.32166: PG=LD-PG: DR=TP*SIN(FN RA(PG))
2390 LD=LD+DR: TT=LD: GOSUB 2470: LD=TT
2400 RM=LD/15: TT=RM: MA=24: GOSUB 2240: RM=TT
2410 AL=LE-NI*.054953: TT=AL: GOSUB2470: AL=TT
2420 AL=LD-AL: TT=AL: MA=360: GOSUB 2240: AL=TT
2430 HE=5.1453*SIN(FN RA(AL))
2440 DM=HE+23.4412*SIN(FN RA(LD))
2450 GOSUB 2180
2460 RETURN
2470 IF TT<-3600 THEN TT=TT+3600: GOTO 2470
2480 IF TT<0 THEN TT=TT+360: GOTO 2480
2490 IF TT>3600 THEN TT=TT-3600: GOTO 2490
2500 IF TT>360 THEN TT=TT-360: GOTO 2500
2510 RETURN
2520 REM
2530 REM *** FIND EQUINOX AND SOLSTICE FOR YEAR ***
2540 REM
2550 PRINT "□▣▣FIND START OF SEASONS."
2560 PRINT "▣ENTER YEAR AS YYYY? ■";
2570 INPUT#1,YR$: PRINT "▣▣▣": YR=VAL(YR$)
2580 FOR CT=3 TO 12 STEP 3: FL=1: MO$=STR$(CT): GOSUB 2730: FL=0
2590 FL=1: GOSUB 2730: FL=0: TM=TM-TZ: IF MO>4 AND MO<11 THEN TM=TM+SD
2600 IF TM<0 THEN TM=TM+24: DA=DA-1
2610 Z=TM: IF Z>=12.98 THEN Z=Z-12
2620 SF$=" A.M.": IF TM>=12 THEN SF$=" P.M."
```

```
2630 GOSUB 3480
2640 PRINT W$(SB);" BEGINS ";MO$(CT);DA;"AT ";V$;SF$;"[illegible]"
2650 NEXT CT
2660 PRINT "[illegible]ANOTHER YEAR? ";
2670 GOSUB 1130: IF Q$="Y" THEN PRINT YS$: GOTO 2550
2680 IF Q$<>"N" THEN 2670
2690 PRINT NO$: GOTO 900
2700 REM
2710 REM *** SUBROUTINE FOR EQUINOX AND SOLSTICE TIMES ***
2720 REM
2730 YR=VAL(YR$): MO=VAL(MO$): GOSUB 3080
2740 SB=INT((MO/3)+.1)-1
2750 T=(365.2422*(YR+LO(SB)/360)-693878.7)/36525
2760 PE=.00134*FN CS(22518.7541*T+153)+.00154*FN CS(45037.5082*T+217)
2770 PE=PE+.002*FN CS(32964.3577*T+313)+.00178*FN SN(20.2*T+231)
2780 M=358.476+35999.04975*T-.00015*T*T-.0000033*T*T*T
2790 L=279.6967+36000.76892*T+.0003025*T*T
2800 L=L+(1.91946-.004789*T-.000014*T*T)*FN SN(M)+(.020094-.0001*T)*FN SN(2*M)
2810 L=L+.000293*FN SN(3*M)-.00479*FN SN(FN NM(259.18-1934.142*T))
2820 L=L-.00569+PE+.00179*FN SN(351+445267.1142*T-.00144*T*T)
2830 L=FN NM(L)
2840 DT=(FN NM(LO(SB)-L+180)-180)/36525
2850 T=T+DT: IF ABS(DT*36525)>.001 THEN 2780
2860 JD=T*36525+2415020.5-(.41+1.2053*T+.4992*T*T)/1440
2870 TM=JD-INT(JD): JD=INT(JD)
2880 TM=TM*24: A$=STR$(INT(60*(TM-INT(TM))+.5))
2890 IF LEFT$(A$,1)=" " THEN A$=MID$(A$,2)
2900 IF LEN(A$)<2 THEN A$="0"+A$: GOTO 2900
2910 REM
2920 REM *** CONVERT JD TO GREG MO/DY/YR ***
2930 REM
2940 C0=INT((JD-1867216.25)/36524.25)
2950 C0=JD+1+C0-INT(C0/4)
2960 C0=C0+1524
2970 C1=INT((C0-122.1)/365.25)
2980 C2=INT(365.25*C1)
2990 C3=INT((C0-C2)/30.6001)
3000 DA=C0-C2-INT(30.61*C3): IF FL=1 THEN RETURN
3010 YR=C1-4716
3020 MO=C3-1
3030 IF MO>12 THEN MO=MO-12: YR=YR-1: GOTO 3030
3040 RETURN
3050 REM
3060 REM *** CONVERT GREG MO/DY/YR TO JD ***
3070 REM
3080 C0=MO: C1=YR: IF MO<3 THEN C0=MO+12: C1=YR-1
3090 JD=INT(365.25*C1)+INT(30.61*(C0+1))+DY+1720995
3100 JD=JD+2-INT(C1/100)+INT(C1/400)
3110 RETURN
3120 REM
3130 REM *** CALCULATE SUNRISE AND SUNSET TIMES ***
3140 REM
3150 FOR I=0 TO 1: IF I THEN J=1.5*π : GOTO 3170
3160 J=TP/4
3170 R=-.0145439: GOSUB 3190: IF I=0 THEN R$=V$: RF$=" A.M."
3180 NEXT I: SF$=" P.M.": RETURN
3190 K=INT((MO+9)/12): X=YR/4: Y=INT(X): Z=X-Y: G=TZ
3200 IF (MO>4 AND MO<11) AND SD=1 THEN G=G-1
3210 G=G*.261799
3220 IF Z=0 THEN 3240
3230 K=K*2
3240 H=INT(275*MO/9)
3250 H=H+DY-K-30: E=LA*.0174533: F=LO*.0174533
3260 K=H+((J+F)/(TP)): L=(K*.017202)-.0574039
3270 Z=SIN(L):M=L+.0334405*Z
3280 Z=SIN(2*L): M=M+.000349066*Z: M=M+4.93289
```

```
3290 Z=M: GOSUB 3540
3300 M=Z: X=M/(TP/4): Y=INT(X): Z=X-Y: IF Z<>0 THEN 3320
3310 M=M+.00000484814
3320 N=2: IF M>1.5*π THEN 3350
3330 N=1: IF M>TP/4 THEN 3350
3340 N=0
3350 P=SIN(M)/COS(M): P=ATN(.91746*P): IF N=0 THEN 3390
3360 IF N=2 THEN 3380
3370 P=P+π: GOTO 3390
3380 P=P+TP
3390 Q=.39782*SIN(M): Q=Q/SQR(-Q*Q+1): Q=ATN(Q)
3400 S=R-(SIN(Q)*SIN(E))
3410 S=S/(COS(Q)*COS(E))
3420 Z=ABS(S): IF Z=<1 THEN 3440
3430 V=0: V$="0:00": RETURN
3440 S=S/SQR(-S*S+1): S=-ATN(S)+TP/4: IF I=1 THEN 3460
3450 S=(TP)-S
3460 Z=.0172028*K: T=S+P-Z-1.73364: U=T+F: V=U-G: Z=V: GOSUB 3540
3470 Z=Z*3.81972
3480 V=INT(Z): W=(Z-V)*60: X=INT(W+.5)
3490 IF X<60 THEN 3510
3500 V=V+1: X=X-60: GOTO 3490
3510 V$=STR$(V+(12*(I=1))): V$=RIGHT$(V$,LEN(V$)-1):IF V$="0" THEN V$="12"
3520 X$=STR$(X): X$=RIGHT$(X$,LEN(X$)-1): IF LEN(X$)=1 THEN X$="0"+X$
3530 V$=V$+":"+X$: V=V+X/100: RETURN
3540 IF Z>=0 THEN 3560
3550 Z=Z+(TP): GOTO 3540
3560 IF Z<(TP) THEN RETURN
3570 Z=Z-(TP): GOTO 3560
3580 REM
3590 REM *** CALCULATE MEAN AGE OF THE MOON ***
3600 REM
3610 YR=VAL(YR$): MO=VAL(MO$): GOSUB 3080: UT=0
3620 T=(JD-2415020+UT)/36525:ET=UT+(.41+1.2053*T+.4992*T*T)/1440
3630 T=(JD-2415020+ET)/36525
3640 L=FN NM(279.6967+36000.7689*T+.000303*T*T)
3650 M=FN NM(358.475833+35999.0498*T-.00015*T*T-3.3E-6*T*T*T)
3660 MP=FN NM(296.104608+477198.8491*T+.009192*T*T+1.44E-5*T*T*T)
3670 F=FN NM(11.250889+483202.0251*T-.003211*T*T-3E-7*T*T*T)
3680 D=JD-2415020: TH=INT(D/10000+.5): D=D-TH*10000
3690 LL=FN NM(270.43416+3.96527*TH+13.1763965*(D+ET)-.001133*T*T+1.9E-6*T*T*T)
3700 OM=LL-F: D=LL-L
3710 X=FN SN(51.2+20.2*T)
3720 LL=LL+.000233*X: M=M-.001778*X: MP=MP+.000817*X
3730 D=D+.002011*X: L=L-.00178*X
3740 X=.003964*FN SN(346.56+132.87*T-.0091731*T*T)
3750 LL=LL+X: MP=MP+X: D=D+X: F=F+X
3760 X=FN SN(OM): LL=LL+.001964*X: MP=MP+.002541*X
3770 D=D+.001964*X
3780 AG=FN NM(D)/360*MR
3790 RETURN
3800 REM
3810 REM *** FIND PHASE OF THE MOON ***
3820 REM
3830 P$=" ": WP$="
3840 IF AG>=28.7 OR AG=<.8 THEN P$="@": WP$="NEW MOON.": RETURN
3850 IF AG>=6.5 AND AG=<8.2 THEN P$="(": WP$="FIRST QUARTER.":RETURN
3860 IF AG>=14.0 AND AG=<15.6 THEN P$="O": WP$="FULL MOON.": RETURN
3870 IF AG>=21.4 AND AG=<23.1 THEN P$=")": WP$="LAST QUARTER."
3880 RETURN
3890 REM
3900 REM *** MONTHLY DATA CALENDAR FLAG SET ***
3910 REM
3920 EV=1: PRINT "[illegible]": GOTO 4000
3930 REM
3940 REM *** MONTHLY BLANK CALENDAR FLAG SET ***
```

```
3950 REM
3960 EV=0: PRINT "▯▣▣▣"
3970 REM
3980 REM *** OUTPUT PROPER MONTHLY CALENDAR ***
3990 REM
4000 PRINT "▣ENTER MONTH, YEAR AS MM,YYYY?": PRINT "? ■";
4010 INPUT#1,MO$,YR$: PRINT "▣"
4020 MO=VAL(MO$): YR=VAL(YR$): DY$="1": DY=1: GOSUB 1190: IF ER=1 THEN 4000
4030 IF EV=1 THEN 4190
4040 PRINT "▣P■RINTER OR ▣S■CREEN?"
4050 GOSUB 1130: IF Q$="P" THEN 4180
4060 IF Q$="S" THEN 4080
4070 GOTO 4050
4080 PRINT "▯■■■■■■■|";MO$(MO);", ";YR$
4090 PRINT "▣SUN  MON  TUE  WED  THU  FRI  SAT"
4100 IF D5=1 THEN 4120
4110 FOR I=1 TO D5-1: PRINT "     ";: NEXT I
4120 CT=D5-1: FOR I=1 TO MO(MO): CT=CT+1
4130 I$=" "+STR$(I): I$=RIGHT$(I$,2)
4140 PRINT " ";I$;"  ";
4150 IF CT/7=INT(CT/7+.5) THEN PRINT "▣"
4160 NEXT I: IF MO(MO)=28 AND LP=1 THEN PRINT " 29  ";
4170 PRINT "▣": GOSUB 1120: GOTO 900
4180 IF EV=0 THEN 4520
4190 FOR I=1 TO 31: EV$(I,4)="": NEXT I
4200 PRINT "▯": FOR CI=1 TO MO(MO)-LP*(MO=2): DY=CI: DY$=STR$(CI)
4210 PRINT "▣▣▣▣▣▣CALCULATING EVENTS FOR ";MO$(MO);CI
4220 GOSUB 3610: T(CI)=AG: GOSUB 3830: P$(CI)=P$
4230 GOSUB 3150: EV$(CI,0)=LEFT$(R$+RF$+BL$,10)
4240 EV$(CI,1)=LEFT$(V$+SF$+BL$,10)
4250 GOSUB 2280: Z=TR: IF Z>=12.98 THEN Z=Z-12
4260 RF$=" A.M.": IF TR>=12 THEN RF$=" P.M."
4270 GOSUB 3480: EV$(CI,2)=LEFT$(V$+RF$+BL$,10)
4280 Z=TS: IF Z>=12.98 THEN Z=Z-12
4290 SF$=" A.M.": IF TS>=12 THEN SF$=" P.M."
4300 GOSUB 3480: EV$(CI,3)=LEFT$(V$+SF$+BL$,10)
4310 IF EV$(CI,4)="" AND CI<>20 THEN EV$(CI,4)=LEFT$(BL$,10): GOTO 4390
4320 IF MO/3<>INT(MO/3) THEN 4390
4330 FL=1: GOSUB 2730: FL=0: TM=TM-TZ: IF MO>4 AND MO<11 THEN TM=TM+SD
4340 IF TM<0 THEN TM=TM+24: DA=DA-1
4350 Z=TM: IF Z>=12.98 THEN Z=Z-12
4360 SF$=" A.M.": IF TM>=12 THEN SF$=" P.M."
4370 GOSUB 3480: EV$(DA,4)=LEFT$(V$+SF$+BL$,10)
4380 IF DA<>20 THEN EV$(CI,4)=LEFT$(BL$,10)
4390 NEXT CI
4400 FOR I=1 TO MO(MO)-LP*(MO=2): IF P$(I)=" " THEN 4510
4410 IF P$(I)<>P$(I+1) THEN 4510
4420 T1=T(I): T2=T(I+1): IF P$(I)="@" THEN T=0
4430 IF P$(I)="(" THEN T=.25
4440 IF P$(I)="O" THEN T=.50
4450 IF P$(I)=")" THEN T=.75
4460 T=MR*T: T1=ABS(T1-T): T2=ABS(T2-T)
4470 IF T1>28 THEN T1=ABS(MR-T(I))
4480 IF T2>28 THEN T2=ABS(MR-T(I+1))
4490 IF T1>T2 THEN P$(I)=" ": GOTO 4510
4500 P$(I+1)=" ": I=I+1: IF P$(I+1)=P$(I-1) THEN P$(I+1)=" ":I=I+1
4510 NEXT I: PRINT "▯▣▣▣OK, PRINTING . . ."
4520 OPEN 4,4: IF EV=0 THEN 4580
4530 T$="ASTRONOMICAL CALENDAR FOR "+MO$(MO)+", "+YR$
4540 PRINT#4,LEFT$(BL$,(80-LEN(T$))/2);T$
4550 T$="PREPARED FOR "+NM$+" AT "+LC$
4560 PRINT#4,LEFT$(BL$,(80-LEN(T$))/2);T$
4570 PRINT#4: PRINT#4: GOTO 4600
4580 PRINT#4: PRINT#4,"                    "MO$(MO)", "YR$: PRINT#4: PRINT#4
4590 FOR II=1 TO 7: FF(II)=-1: NEXT II
4600 GOSUB 5010: GOSUB 4990
```

```
4610 PRINT#4,"!  "DY$(1)"  !  "DY$(2)"  !  "DY$(3)" ! "DY$(4)"!";
4620 PRINT#4," "DY$(5)" !  "DY$(6)"  ! "DY$(7)" !": GOSUB 4990: GOSUB 5030
4630 PRINT#4,"!";: IF D5=1 THEN 4650
4640 FOR I=1 TO D5-1: PRINT#4,"         !";: NEXT I
4650 CT=D5-1: FOR I=1 TO MO(MO): CT=CT+1: FF(CT)=I
4660 I$=" "+STR$(I): I$=RIGHT$(I$,2): PRINT#4,"  ";
4670 IF EV=1 THEN PRINT#4,P$(I);"     "I$" !";: GOTO 4690
4680 PRINT#4,"       "I$" !";
4690 IF CT=7 THEN PRINT#4
4700 IF CT=7 THEN PRINT#4,"!";: FOR II=1 TO 7: PRINT#4,"         !";: NEXT II
4710 IF CT=7 THEN PRINT#4
4720 IF EV=1 AND CT=7 THEN GOSUB 5080
4730 IF EV=0 AND CT=7 THEN GOSUB 4960
4740 IF CT=7 THEN GOSUB 5030
4750 IF CT=7 AND I<MO(MO) THEN CT=0: PRINT#4,"!";
4760 IF CT=7 THEN CT=0
4770 NEXT I: IF MO(MO)=28 AND LP=1 THEN PRINT#4,"      29 !";: CT=CT+1
4780 IF CT=7 OR CT=0 THEN CLOSE 4: GOTO 900
4790 FOR II=CT+1 TO 7: PRINT#4,"         !";: NEXT II: PRINT#4
4800 PRINT#4,"!";: FOR II=1 TO 7: PRINT#4,"         !";: NEXT II: PRINT#4
4810 IF EV=0 THEN GOSUB 4960
4820 IF EV=1 AND (CT>1 AND CT<7) THEN GOSUB 5080
4830 GOSUB 5030: IF EV=0 THEN CLOSE 4: GOTO 900
4840 PRINT#4
4850 PRINT#4,"@ = NEW MOON, ( = FIRST QUARTER, ";
4860 PRINT#4,"O = FULL MOON, ) = LAST QUARTER."
4870 PRINT#4,"SUNRISE, SUNSET, MOONRISE, MOONSET";
4880 IF MO/3=INT(MO/3) THEN PRINT#4,", "W$(MO/3-1);
4890 IF MO/6=INT(MO/6) THEN PRINT#4," SOLSTICE";: GOTO 4910
4900 IF MO/3=INT(MO/3) THEN PRINT#4," EQUINOX";
4910 PRINT#4,"."
4920 CLOSE 4: GOTO 900
4930 REM
4940 REM *** SUBROUTINE PRINT BLANK CALENDAR ***
4950 REM
4960 FOR J=1 TO 6: PRINT#4,"!";
4970 FOR II=0 TO 4: PRINT#4,"         !";: NEXT II: PRINT#4
4980 NEXT J
4990 PRINT#4,"!";: FOR J=0 TO 6: PRINT#4,"         !";: NEXT J
5000 PRINT#4: RETURN
5010 PRINT#4,"------------------------------------------------------------------";
5020 PRINT#4,"-----------------------": RETURN
5030 PRINT#4,"+";: FOR J=0 TO 6: PRINT#4,"----------+";: NEXT J
5040 PRINT#4: RETURN
5050 REM
5060 REM *** SUBROUTINE TO PRINT EVENTS CALENDAR ***
5070 REM
5080 FOR J=0 TO 4: PRINT#4,"!";: FOR II=1 TO 7
5090 IF FF(II)=-1 THEN 5110
5100 PRINT#4,LEFT$(EV$(FF(II),J)+BL$,10);"!";: GOTO5120
5110 PRINT#4,"          !";
5120 NEXT II: PRINT#4: NEXT J
5130 FOR II=1 TO 7: FF(II)=-1: NEXT II
5140 RETURN
5150 REM
5160 REM *** ROUTINE TO READ USER DATA ***
5170 REM
5180 OPEN 2,8,3,"0:NAMDAT"
5190 INPUT#15,A1$,B$,C$,D$: IF A1$="62" THEN CLOSE 2: GOTO 5280
5200 IF A1$<>"00" THEN 5640
5210 INPUT#2,NM$: INPUT#2,LC$: INPUT#2,LA$
5220 INPUT#2,LO$: INPUT#2,TZ$: INPUT#2,SD: TZ=VAL(TZ$): CLOSE 2: RETURN
5230 REM
5240 REM *** SUBROUTINE TO INPUT USER DATA ***
5250 REM
5260 ED=1: POKE 53280,6: POKE 53281,14
```

```
5270 PRINT "■■■     EDIT PERSONAL USE DATA": GOSUB 5290: GOTO 900
5280 POKE 53280,11: POKE 53281,1: PRINT "■■■     ENTER PERSONAL USE DATA"
5290 PRINT "■■PLEASE ENTER YOUR NAME? ■";NM$;"      ";LEFT$(CL$,5+LEN(NM$));
5300 INPUT#1,NM$: PRINT "■"
5310 PRINT "■WHAT IS YOUR LOCATION"
5320 PRINT "? ■";LC$;"      ";LEFT$(CL$,5+LEN(LC$));: INPUT#1,LC$: PRINT "■"
5330 PRINT "■WHAT IS THE LATITUDE"
5340 PRINT "OF ";LC$;"? ■";LA$;"       ";LEFT$(CL$,6+LEN(LA$));
5350 INPUT#1,LA$: PRINT "■"
5360 LA=VAL(LA$): IF LA<-90 OR LA>90 THEN PRINT "■■": GOTO 5330
5370 PRINT "■WHAT IS THE LONGITUDE"
5380 PRINT "OF ";LC$;"? ■";LO$;"       ";LEFT$(CL$,6+LEN(LO$));
5390 INPUT#1,LO$: PRINT "■"
5400 LO=VAL(LO$): IF LO<0 OR LO>359 THEN PRINT "■■": GOTO 5370
5410 PRINT "■IN WHICH TIME ZONE IS"
5420 PRINT LC$;" (0-23)? ■";TZ$;"    ";LEFT$(CL$,3+LEN(TZ$));
5430 INPUT#1,TZ$: PRINT "■"
5440 TZ=VAL(TZ$): IF TZ<0 OR TZ>23 THEN PRINT "■■": GOTO 5410
5450 PRINT "■IS DAYLIGHT SAVING TIME USED IN ": PRINT LC$;"?     ■■■■";
5460 GOSUB 1130: IF Q$="Y" THEN PRINT YS$: SD=1: GOTO 5490
5470 IF Q$<>"N" THEN 5460
5480 SD=0: PRINT NO$
5490 PRINT "■IS THIS CORRECT?     ■■■■";
5500 GOSUB 1130: IF Q$="Y" THEN PRINT YS$: GOTO 5550
5510 PRINT NO$;"■■■": GOTO 5290
5520 REM
5530 REM *** WRITE USER DATA FILE ***
5540 REM
5550 OPEN 2,8,3,"0:NAMDAT,S,W"
5560 INPUT#15,A1$,B$,C$,D$: IF A1$="63" AND ED=1 THEN 5630
5570 IF A1$<>"00" THEN 5640
5580 PRINT#2,NM$: PRINT#2,LC$: PRINT#2,LA$
5590 PRINT#2,LO$: PRINT#2,TZ$: PRINT#2,SD: CLOSE 2: RETURN
5600 REM
5610 REM *** HANDLE DISK ERROR ***
5620 REM
5630 CLOSE 2: PRINT#15,"S0:NAMDAT": GOTO 5550
5640 TR=0: CLOSE 2: PRINT "■DISK ERROR -": PRINT A1$;" ,";B$;",";C$;",";D$
5650 PRINT "■TRY AGAIN? ";
5660 GOSUB 1130: IF Q$="Y" THEN PRINT YS$: TR=1: RETURN
5670 IF Q$="N" THEN TR=0: PRINT NO$: RETURN
5680 GOTO 5660
5690 REM
5700 REM *** VIDEO DISPLAY OF EVENTS FOR A DAY ***
5710 REM
5720 PRINT "■■■■■"
5730 PRINT "■ENTER MONTH, DAY, YEAR AS MM,DY,YYYY?": PRINT "? ■";
5740 INPUT#1,MO$,DY$,YR$: PRINT "■"
5750 MO=VAL(MO$): YR=VAL(YR$): DY=VAL(DY$): GOSUB 1190: IF ER=1 THEN 5730
5760 MO=VAL(MO$): PRINT "■■DATA FOR ";DY$(N);" ";MO$(MO);" ";DY$;",  ";YR$
5770 PRINT "■■";TAB(16);"RISE          SET"
5780 PRINT TAB(13);"■==========    =========="
5790 PRINT "■SUN■";: GOSUB 3150
5800 PRINT TAB(15+(LEN(R$)>4));R$;RF$;TAB(27+(LEN(V$)>4));V$;SF$
5810 PRINT "■MOON■";: GOSUB 2280: Z=TR: IF Z>=12.98 THEN Z=Z-12
5820 RF$=" A.M.": IF TR>=12 THEN RF$=" P.M."
5830 GOSUB 3480: PRINT TAB(15+(LEN(V$)>4));V$;RF$;
5840 Z=TS: IF Z>=12.98 THEN Z=Z-12
5850 SF$=" A.M.": IF TS>=12 THEN SF$=" P.M."
5860 GOSUB 3480: PRINT TAB(27+(LEN(V$)>4));V$;SF$
5870 IF MO/3<>INT(MO/3) THEN 5980
5880 IF DY>19 AND DY<24 THEN FL=1: GOSUB 2730: FL=0: GOTO 5900
5890 GOTO 5980
5900 IF DA<>DY THEN 5980
5910 TM=TM-TZ: IF MO>4 AND MO<11 THEN TM=TM+SD
5920 IF TM<0 THEN TM=TM+24: DA=DA-1
```

```
5930 IF DA<>DY THEN 5980
5940 Z=TM: IF Z>=12.98 THEN Z=Z-12
5950 SF$=" A.M.": IF TM>=12 THEN SF$=" P.M."
5960 GOSUB 3480
5970 PRINT "■■";W$(SB);" BEGINS AT ■";V$;SF$;"■■"
5980 GOSUB 3610: GOSUB 3830
5990 AG=INT(AG*100+.5)/100
6000 IF AG=0 THEN T$="0.00": GOTO 6060
6010 T$=STR$(AG): T$=RIGHT$(T$,LEN(T$)-1)
6020 IF AG=INT(AG) THEN T$=T$+"."
6030 IF AG<1 THEN T$="0"+T$
6040 IF AG<10 THEN T$=LEFT$(T$+"0000",4)
6050 IF AG>=10 THEN T$=LEFT$(T$+"0000",5)
6060 PRINT "■■THE MEAN AGE OF THE MOON IS ■";T$;"■ DAYS."
6070 IF WP$<>"" THEN PRINT "■IT IS ";WP$
6080 GOSUB 1470
6090 PRINT "■■ANOTHER DAY? ";
6100 GOSUB 1130: IF Q$="Y" THEN PRINT YS$: GOTO 5720
6110 IF Q$<>"N" THEN 6100
6120 PRINT NO$: GOTO 900
6130 REM
6140 REM *** NORMAL EXIT ***
6150 REM
6160 POKE 53280,14: POKE 53281,6
6170 POKE 808,237: PRINT "■■";CHR$(9): END

READY.
```

Table F-1. Variable Descriptions for the Astronomical Calendar.

A —Menu selection variable.
A$ —Minutes portion of equinox and solstice times.
A1$ —Disk error number.
AG —Age of the moon for current lunar cycle.
AL —Used to calculate moonrise/moonset.
B —Temporary in Easter calculation.
B$ —Disk error message.
BL$ —String with 40 spaces.
C$ —Disk error track number.
C0 —Used in Gregorian date to Julian day conversions.
C1 —Used in Gregorian date to Julian day conversions.
C2 —Used in Gregorian date to Julian day conversions.
C3 —Used in Gregorian date to Julian day conversions.
CD$ —String with home and 26 cursor down characters.
CI —Loop counter.
CL$ —String with 48 cursor left characters.
CR$ —String with 48 cursor right characters.
CT —Loop counter.
D —Numeric day number.
D$ —Disk error sector number.
D5 —Number of day of the week.
DA —Calculated day of the month.
DD —Numeric day number.
DE —Day number for Easter.
DG —Days from January 1 to date.
DM —Declination of moon on date.
DR —Used in moonrise/moonset calculations.
DT —Intermediate in solstice and equinox calculations.
DY —Day number.
DY$ —String variable with day number.
DY$(—Array with names for days of the week.

E —Used in sunrise/sunset calculations.
ED —Calculated date for Easter.
EP —Number of days from 0,0,0 to 1,1,1960
ER —Flag for error.
ET —Ephimeris time for date at Oh UT.
EV —Flag for blank or events calendar.
EV$(—Daily events for printing events calendar.
F —Used in sunrise/sunset calculations.
FF(—Controls printing for week in events calendar.
FL —Flag to control conversion of Julian day to month/year.
G —Used in sunrise/sunset calculations.
GC —Greenwich Mean Time at Oh on January 0.
H —Used in sunrise/sunset calculations.
HD$ —Holiday name.
HD$(—Holiday name array.
HD(—Holiday day of the month array.
HE —Current inclination of moon in its orbit.
HH —Current number of holidays.
HM(—Holiday month number array.
I —Loop counter.
I$ —String of day of the month in calendar printing.
II —Loop counter.
J —Loop counter.
JD —Julian day equivalent to date.
JJ —Flag for match for holiday.
K —Used in calculation of sunrise/sunset.
L —Temporary numeric variable.
L1 —Normalized value of user's longitude.
LA —User's latitude.
LA$ —String with user's latitude.
LC$ —Name of user's location.
LD —Used in calculation of moonrise/moonset.
LE —Constant used in calculation of moonrise/moonset.
LL —Used in calculation of mean age of the moon.
LO —User's longitude.
LO —String with user's longitude.
LO(—Array for longitude of sun at solstices or equinoxes.
LP —Flag for leap year.
LZ —Constant used in calculation of moonrise/moonset.
M —Numeric value of month.
MA —Maximum value used for modulo function.
ME —Month when Easter occurs.
MM —Numeric value of month.
MO —Numeric value of month.
MO$—String with numeric value of month.
MO$(—Array with names of the months.
MO(—Array with number of days per month.
MP —Used to calculate mean age of the moon.
MR —Number of days in lunar month.
N —Temporary numeric variable.
NI —Number of days from Jan. 0, 1960.
NM$—User's name.
NO$—Constant string with the word NO.
OM —Longitude of the lunar ascending mode.
P —Used in calculation of sunrise/sunset.
P$ —Symbol for phase of the moon.
P$(—Array with symbols for phases of the moon for a month.
PE —Used in mean age of the moon calculations.
PG —Used in moonrise/moonset calculations.
PL —Used in moonrise/moonset calculations.
Q —Temporary numeric variable.

Q$ —Keyboard input variable.
R —Used in sunrise/sunset calculations.
R$ —String sunrise time.
RD —Radian-degree conversion factor.
RF$ —String for A.M./P.M. notations for rising phenomena.
RM —Right ascension for the moon.
S —Used in sunrise/sunset calculations.
SB —Quarter for solstice/equinox calculations.
SD —Flag for use of daylight savings time.
SF$ —String for A.M./P.M. notations for setting phenomena.
T —Temporary numeric variable.
T$ —Temporary string variable.
T(—Array of mean age of the moon for calendar printout.
T1 —Temporary numeric variable.
T2 —Temporary numeric variable.
TA —Temporary numeric variable.
TC —Constant for daily change in sidereal time.
TH —Used in mean age of the moon calculations.
TM —Calculated time variable.
TP —Constant with the value of two time pi.
TR —Time for moon rise.
TS —Time for moon set.
TT —Temporary numeric variable.
TZ —User's time zone.
TZ$ —String with user's time zone.
U —Temporary numeric variable.
UT —Universal time.
V —Temporary numeric variable.
V$ —String sunset, moonrise, moonset time.
W —Temporary numeric variable.
W$(—Array with season names.
WP$—Word description of the phase of the moon.
X —Temporary numeric variable.
X$ —Temporary string variable used in conversion of decimal time to sexigesimal time notation.
Y —Temporary numeric variable.
YD —Number of days since January 1.
YR —Numeric value of year.
YR$ —String with value of the year.
YS$ —String with the word YES.
YY —Numeric value of year.
Z —Temporary numeric variable.

Following is a discussion of each of these options in the order that they appear on the program menu:

Print Monthly Calendar

Pressing the number 1, when the program menu is displayed, will select the Print monthly calendar menu option. After you press the number 1, the screen will change colors and you will be asked to:

ENTER MONTH, YEAR AS MM,YYYY?

This means that you should enter the number of the month and year. If, for example, you are wanting a calendar of the month of

November, 1986, enter the data as 11,1986. The computer will then ask:

PRINTER OR SCREEN?

Note that the first letter of the words "printer" and "screen" are displayed in reverse characters. The reverse characters are to indicate which keys you are to press to make your choice of which option you want to use. To have the calendar printed on your printer, press the P key. To have the calendar displayed on the screen, press the S key.

The calendar, when displayed on the screen, will appear like that in Fig. F-2, while a printed calendar will appear like that in Fig. F-3.

Fig. F-2. Monthly calendar.

NOVEMBER, 1986

SUNDAY	MONDAY	TUESDAY	WEDNESDAY	THURSDAY	FRIDAY	SATURDAY
						1
2	3	4	5	6	7	8
9	10	11	12	13	14	15
16	17	18	19	20	21	22
23	24	25	26	27	28	29
30						

Fig. F-3. Celestial Events Calendar Program.

Print Monthly Events Calendar

Pressing the number 2, when the program menu is displayed, will select the Print monthly events calendar option. This option operates much like the previous one, but without a screen option. Because of the number and length of calculations, you can expect that this option will take up to 10 minutes to produce a calendar for a single month. The further the distance in time from the year 2000, the longer the calculations will take. A typical calendar is reproduced in Fig. F-4.

To let you know that the program is working during that delay in printing, the screen will tell you the day for which the computer is calculating events. Most of the time, you will experience a delay during the printing of the calendar. This delay is normal. It means that the computer is doing some internal housekeeping called garbage collection. The computer will continue with the calculations after a delay of 10-15 seconds.

For each day on the calendar, you will have as many as six separate pieces of information. To the left of the date is a symbol for the phase of the moon. The key for these symbols is printed on the bottom of the calendar page. The first time below the date is the sunrise time calculated for your location. This is followed by the sunset time. If neither of these occur at your location on that date, the computer will print 0:00 for the time. This would be for locations in the far north for times when there is neither sunrise or sunset.

Next on the calendar comes the moonrise time, followed by the moonset time. Note that these times are for your location. Finally, on four dates in the year the calendar will print the time for the solstice or equinox—the beginnings of the seasons.

Calculate the Start of the Seasons

This option is selected by pressing the number 3 on the keyboard. The computer will ask that you:

ENTER THE YEAR AS YYYY?

An example of YYYY could be 1986. After entering the year and pressing RETURN, there will be a brief pause, then the computer will display the start of each of the seasons for the year selected.

Display a Day's Events

You can select this option from the program menu by pressing the number 4 on the keyboard. The computer will ask you for the date:

ENTER THE DATE AS MM,DD,YYYY

ASTRONOMICAL CALENDAR FOR MARCH, 1987
PREPARED FOR DICK DALEY AT DARBY

SUNDAY	MONDAY	TUESDAY	WEDNESDAY	THURSDAY	FRIDAY	SATURDAY
1 7:08 A.M. 6:15 P.M. 7:28 A.M. 7:34 P.M.	2 7:06 A.M. 6:16 P.M. 7:50 A.M. 8:51 P.M.	3 7:04 A.M. 6:18 P.M. 8:11 A.M. 10:04 P.M.	4 7:02 A.M. 6:19 P.M. 8:33 A.M. 11:15 P.M.	5 7:01 A.M. 6:21 P.M. 8:57 A.M. 12:22 A.M.	6 6:59 A.M. 6:22 P.M. 9:24 A.M. 1:24 A.M.	(7 6:57 A.M. 6:23 P.M. 9:56 A.M. 2:19 A.M.
8 6:55 A.M. 6:25 P.M. 10:34 A.M. 3:07 A.M.	9 6:53 A.M. 6:26 P.M. 11:20 A.M. 3:47 A.M.	10 6:51 A.M. 6:27 P.M. 12:14 P.M. 4:19 A.M.	11 6:49 A.M. 6:29 P.M. 1:14 P.M. 4:47 A.M.	12 6:47 A.M. 6:30 P.M. 2:18 P.M. 5:10 A.M.	13 6:45 A.M. 6:32 P.M. 3:26 P.M. 5:32 A.M.	O 14 6:44 A.M. 6:33 P.M. 4:38 P.M. 5:53 A.M.
15 6:42 A.M. 6:34 P.M. 5:51 P.M. 6:14 A.M.	16 6:40 A.M. 6:36 P.M. 7:07 P.M. 6:35 A.M.	17 6:38 A.M. 6:37 P.M. 8:25 P.M. 6:57 A.M.	18 6:36 A.M. 6:38 P.M. 9:43 P.M. 7:21 A.M.	19 6:34 A.M. 6:40 P.M. 11:01 P.M. 7:47 A.M.	20 6:32 A.M. 6:41 P.M. 12:17 A.M. 8:18 A.M. 8:53 P.M.	21 6:30 A.M. 6:42 P.M. 1:25 A.M. 8:57 A.M.
) 22 6:28 A.M. 6:44 P.M. 2:24 A.M. 9:47 A.M.	23 6:26 A.M. 6:45 P.M. 3:11 A.M. 10:49 A.M.	24 6:24 A.M. 6:46 P.M. 3:47 A.M. 12:01 P.M.	25 6:22 A.M. 6:48 P.M. 4:17 A.M. 1:18 P.M.	26 6:20 A.M. 6:49 P.M. 4:43 A.M. 2:39 P.M.	27 6:18 A.M. 6:50 P.M. 5:07 A.M. 3:59 P.M.	28 6:16 A.M. 6:52 P.M. 5:29 A.M. 5:18 P.M.
@ 29 6:15 A.M. 6:53 P.M. 5:55 A.M. 6:39 P.M.	30 6:13 A.M. 6:54 P.M. 6:17 A.M. 7:54 P.M.	31 6:11 A.M. 6:56 P.M. 6:39 A.M. 9:07 P.M.				

@ = NEW MOON, (= FIRST QUARTER, O = FULL MOON,) = LAST QUARTER.
SUNRISE, SUNSET, MOONRISE, MOONSET, SPRING EQUINOX.

Fig. F-4. Astronomical calendar for March, 1987 prepared for Dick Daley at Darby.

This means that July 4, 1989 would be entered as 07,04,1989.

This option will display on the screen the same information that is shown in each frame of the Monthly Events Calendar. In addition, a value for the mean age of the moon is shown. When the number is zero, it is a new moon. As this value increases from zero to about 14.75 the moon is a waxing moon. At the age of 14.75, it is full moon. As the age increases to about 29.5 days, the moon is waning. Then the age cycles back to zero again. While the computer is checking for the phase of the moon, not only does it inform you of the actual phase, it also tells you when it is near full moon. The same is true for the new moon, and the first and last quarters.

Besides the strictly astronomical data, the computer also knows when eighteen different common American holidays occur. If the date you select is one of these, you will be told which of the holidays you have chosen.

Edit Personal Use Data

Select this option from the program menu by pressing the number 5 on the keyboard. Here you can correct errors you may have made on the data for personalizing your calendar, which you entered the first time you used the program. This option works just like the editing did when you first entered data. For questions see the beginning of this Appendix.

Exit

When you have finished using the Astronomical Calendar, it is best to exit the program by using this Exit option. Pressing a 6 on the keyboard while the program menu is displayed will allow you to exit the program.

Index

Other Bestsellers From TAB

☐ **COSMOLOGY: THE SEARCH FOR THE ORDER OF THE UNIVERSE—Caes**

Was there really a "Big Bang" that created our universe? How many stars exist in the universe at any one moment? These are just a few of the many mysteries of our universe that are explored in this tantalizing look at man's ongoing search for an understanding of cosmic order. This is a selection that none interested in astronomy, cosmology, cosmogony, or astrology should miss! 192 pp., 24 illus. 7″ × 10″.

Paper $10.95 **Hard $16.95**
Book No. 2626

☐ **COMETS, METEORS AND ASTEROIDS —How They Affect Earth—Gibilisco**

Information on meteors, asteroids, and other related space phenomenon is all here for the taking. Includes a spectacular eight page section of color photos taken in space. Packed with little-known details and fascinating theories covering history's most memorable comets—including Halley's Comet, the origins of the solar system, and speculation on what may happen in the future. 224 pp., 148 illus.

Paper $14.95 **Book No. 1905**

☐ **333 MORE SCIENCE TRICKS AND EXPERIMENTS—Brown**

Here's an ideal way to introduce youngsters of all ages to the wonders and complexities of science . . . a collection of tricks and experiments that can be accomplished with ordinary tools and materials. You can demonstrate that air is "elastic," perform hydrotropism, or psychological tricks . . . or perform any one of more than 300 fascinating experiments. 240 pp., 189 illus.

Paper $10.95 **Hard $15.95**
Book No. 1835

☐ **VIOLENT WEATHER: HURRICANES, TORNADOES AND STORMS—Gibilisco**

What causes violent storms at sea? Hurricane force winds? Hail the size of grapefruit? Blinding snowstorms and tornadoes? The answers to all these and many more questions on the causes, effects, and ways to protect life and property from extremes in weather are here in this thoroughly fascinating study of how extremes in weather violence occur. 272 pp., 192 illus. 7″ × 10″.

Paper $13.95 **Book No. 1805**

☐ **PARTICLES IN NATURE: THE CHRONOLOGICAL DISCOVERY OF THE NEW PHYSICS—Mauldin**

If you're interested in physics, science, astronomy, or natural history, you will find this presentation of the particle view of nature fascinating, informative, and entertaining. John Mauldin has done what few other science writers have been able to accomplish . . . he's reduced the complex concepts of particle physics to understandable terms and ideas. This enlightening guide makes particle physics seem less abstract—it shows significant spin-offs that have resulted from research done, and gives a glimpse of future research that promises to be of practical value to everyone. 304 pp., 169 illus. 16 Full-Color Pages, 14 Pages of Black & White Photos. 7″ × 10″.

Paper $16.95 **Hard $23.95**
Book No. 2616

☐ **TIME GATE: HURTLING BACKWARD THROUGH HISTORY—Pellegrino**

Taking a new approach to time travel, this totally fascinating history of life on Earth transports you backward from today's modern world through the very beginnings of man's existence. Interwoven with stories and anecdotes and illustrated with exceptional drawings and photographs, this is history as it should always have been written! It will have you spellbound from first page to last! 288 pp., 142 illus. 7″ × 10″.

Paper $16.95 **Book No. 1863**

☐ **333 SCIENCE TRICKS AND EXPERIMENTS—Brown**

Here is a delightful collection of experiments and "tricks" that demonstrate a variety of well-known, and not so well-known scientific principles and illusions. Find tricks based on inertia, momentum, and sound projects based on biology, water surface tension, gravity and centrifugal force, heat, and light.Every experiment is easy to understand and construct . . . using ordinary household items. 208 pp., 189 illus.

Paper $9.95 **Book No. 1825**

☐ **THE COMPLETE PASSIVE SOLAR HOME BOOK—Schepp and Hastie**

You'll get down-to-earth pointers on basic energy conservation . . . a clear picture of passive solar home design options—underground homes, superinsulated homes, double envelope houses, and manufactured housing. Plus, you'll get expert how-to's for choosing a passive solar design . . . how to deal with architects, designers, and contractors . . . and more. 320 pp., 252 illus. 7″ × 10″.

Paper $16.95 **Book No. 1657**

4/8